Cosmic perspectives

M.K. Vainu Bappu

Cosmic perspectives

Essays dedicated to the memory of
M.K.V. Bappu

EDITED BY

S.K. BISWAS
Indian Institute of Science
Bangalore, India

D.C.V. MALLIK
Indian Institute of Astrophysics
Bangalore, India

C.V. VISHVESHWARA
Raman Research Institute
Bangalore, India

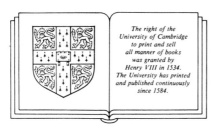

The right of the
University of Cambridge
to print and sell
all manner of books
was granted by
Henry VIII in 1534.
The University has printed
and published continuously
since 1584.

CAMBRIDGE UNIVERSITY PRESS
CAMBRIDGE
NEW YORK   PORT CHESTER
MELBOURNE   SYDNEY

Published by the Press Syndicate of the University of Cambridge
The Pitt Building, Trumpington Street, Cambridge CB2 1RP
40 West 20th Street, New York, NY 10011, USA
10 Stamford Road, Oakleigh, Melbourne 3166, Australia

First published 1989

Printed in Great Britain by Alden Press, Oxford

British Library CIP data

Cosmic perspectives.
1. Astronomy. Cosmology
I. Biswas, S.K., *1945–* II. Mallik, D.C.V. (Dipankar
C.V.), *1946–* III. Vishveshwara, C.V., *1938–*
IV. Bappu, M.K.V.
523.1

Library of Congress CIP data

Cosmic perspectives: essays dedicated to the memory of M.K.V. Bappu
edited by S.K. Biswas, D.C.V. Mallik, and C.V. Vishveshwara.
    p.   cm.
ISBN 0 521 34354 2
1. Astronomy. 2. Cosmology. 3. Astrology. 4. Bappu, M.K.V.
I. Bappu, M.K.V. II. Biswas, S.K. III. Mallik, D.C.V.
IV. Vishveshwara, C.V.
QB51.C74 1989
520–dc19   88–28708CIP

ISBN 0 521 34354 2

AL

Contents

Contributors

R. HANBURY BROWN
School of Physics
University of Sydney
Sydney N.S.W. 2006
Australia

BERNARD CARR
School of Mathematical Sciences
Queen Mary College
University of London
Mile End Road
London E1 4NS
U.K.

BRANDON CARTER
Observatoire de Paris-Meudon
Place J. Janssen
92195-Meudon Principal Cedex
France

DEBIPRASAD CHATTOPADHYAYA
History of Science and Technology
  Project
3 Sambhunath Pandit Street
Calcutta – 700 020
India

ROGER CULVER
Department of Physics
Colorado State University
Fort Collins
Colorado 80523
U.S.A.

FRED HOYLE
Cockley Moore
Dockray
Cumberland CA11 0LG
U.K.

ALLEN I. JANIS
Department of Physics & Astronomy
University of Pittsburgh
Pittsburgh
Pennsylvania 15260
U.S.A.

IVAN W. KELLY
Department of Educational Psychology
University of Saskatchewan
Saskatoon, Sask. S7N 0WO
Canada

PETER J. LOPTSON
Department of Philosophy
University of Saskatchewan
Saskatoon, Sask. S7N 0WO
Canada

JAYANT V. NARLIKAR
Tata Institute of Fundamental Research
Homi Bhabha Road
Bombay – 400 005
India

JOSEPH NEEDHAM
The Needham Research Institute
East Asian History of Science
  Library
16 Brooklands Avenue
Cambridge CB2 2BB
U.K.

JEAN-CLAUDE PECKER
LAT
98 Bis Blvd Arago
F-75014 Paris
France

CYRIL PONNAMPERUMA
Laboratory of Chemical Evolution
University of Maryland
College Park, MD 20742
U.S.A.

and

Institute of Fundamental Studies
Huntane
Kandy
Sri Lanka

HUBERT REEVES
Research Center for Nuclear Studies
SEP-SES BAT 28
Cen Saclay
B P 2
F-91190 Gif s Yvette
Saclay
France

HARLAN SMITH, JR
McDonald Observatory
University of Texas
Austin
Texas 78712
U.S.A.

B.V. SUBBARAYAPPA
Centre for History and Philosophy of
   Science
Indian Institute of World Culture
B.P. Wadia Road
Bangalore – 560 004
India

C.V. VISHVESHWARA
Raman Research Institute
Bangalore – 560 080
India

Preface

The contemplation of the heavens has been one of the most enduring adventures of the human mind. The awe and inspiration this evoked since the earliest of times led naturally to the growth of astronomy and cosmology. They, in turn, helped the development of other sciences including mathematics. Astronomy as it grew permeated every sphere of human activity. On the one hand, it profoundly influenced human thought at the highest level; on the other, it served totally utilitarian purposes as in navigation, time-keeping and agriculture. As it evolved, it revealed an increasingly clear picture of the material world. However, the inevitable gaps in man's knowledge were often sought to be filled in by irrational beliefs leading to myths and religious dogma. Consequently, astronomy has also been a battlefield for the fierce conflict between enlightenment and ignorance. The latter persists to this day in the form of astrology and related pseudoscientific pursuits. Astronomy is thus a force which seeks to liberate man while its perversion keeps millions slaves to superstitions.

Astronomy and cosmology present innumerable fascinating aspects in their impact of human knowledge and their interaction with different areas of culture and civilization. At the same time, they continue their exciting quest for understanding nature. The present volume is an attempt to bring together these diverse facets of astronomy and cosmology.

The need for such a book was felt by the editors and some of their colleagues some time ago while exploring the possibility of bringing out a special issue of the bimonthly journal, *The Bulletin of Sciences*. This journal is edited and published by a group of working scientists and is devoted to discussions of the interaction of science with society. When leading experts in the various fields related to astronomy responded favourably to our request to write, we realised that a book covering such a wide spectrum of topics was indeed possible.

The articles in this volume focus broadly on important areas such as history of astronomy, the interaction of astronomy with society, science and culture, the structure and exploration of the universe, origin of life and man's place in the universe. In cosmology emphasis has been placed on nonstandard approaches for the reason that standard scenarios have been extensively elucidated in other books. The last two articles in the book are a departure from convention. One is a keen analysis of astrology while the other is a delightful journey into the world of fantasy and imagination.

This book is dedicated to the memory of Professor M.K. Vainu Bappu. Vainu Bappu was one of the architects of modern astronomy in India and built institutions devoted to research in astronomy and cosmology. It is our hope that this volume is a fitting tribute to a man whose vision was broad enough to embrace all aspects of astronomy.

We are grateful to all the contributors who so warmly responded to our request to write. We thank the Cambridge University Press for their help and cooperation in bringing out this volume. It is with great pleasure we acknowledge the enthusiastic response of Dr Simon Mitton, Editorial Director, Science, Technology and Medicine, Cambridge University Press, and the subsequent support received from him.

We acknowledge the valuable help of Dr Sabyasachi Chatterjee and the encouragement given by the Government of Karnataka without which the project could not have got off the ground. We thank Dr Margaret Biswas for editorial help. We are grateful to the Directors of the respective Institutes where we work, Professors C.N.R. Rao, J.C. Bhattacharya and V. Radhakrishnan, who supported the project and provided institutional facilities to make it a reality. We also thank Ms Moksha Halesh, Ms R.K. Lakshmi, Ms Meena Krishnan, Ms Meenakshi, Mr S. Rajasekharan and Mr A.M. Batcha for secretarial help, Mr K. Govindan, Mr Raju Verghese and Mr C. Ramachandra Rao for drawing and photographic work and Mr G. Chandramohan for editorial help.

Bangalore   S.K. Biswas
                                                                    D.C.V. Mallik
                                                                C.V. Vishveshwara

Foreword

Science and culture

It is an honour to introduce a book which is dedicated to the memory of a distinguished astronomer and a good friend, the late Professor M.K.V. Bappu. As all of us who knew him soon found out, he was interested not only in the scientific aspects of astronomy but also in its history and relations with society. The essays presented in this book are a fitting tribute to the memory of a man of such wide interests. Vainu Bappu would, I feel sure, have appreciated them.

Astronomy has always played a major part in the development of science. As Tycho Brahe pointed out in the 16th century, to comprehend what we see in the sky we must look at what lies at our feet (*suspiciendo = despicio*]; to comprehend what lies at our feet we must look at the sky [*despiciendo = suspicio*). A glance at modern high-energy physics shows us how true that still is.

But that is not the only way in which astronomy is valuable to science. Astronomy and medical science have always been the principal ambassadors at the court of public opinion. That medical science is useful is transparently obvious to anyone and its credentials are accepted by society without question – there is no branch of science for which it is easier to get public support than research into cancer. The credentials of astronomy on the other hand, are accepted by society for quite a different reason. Although its practical value is not generally recognised, astronomy is seen as something which is worth doing in itself. It affords all of us, no matter how little we know about mathematics, quarks or biochemistry, a glimpse of the wonder and immensity of the mysterious Universe in which we find ourselves. As such it is a major link between science and the man and woman in the street.

If we were to ask the man and woman in the street what they believe to be the role of science in society, their answer would depend on when and where we asked that question. In medieval Europe, if we could have found anyone who had ever heard of science, they would have told us that an understanding of the natural world is something which we inherit from the Ancient Greeks. Don't bother with scientific research, they would have said, just go into the library and read the works of Aristotle, Galen and St Thomas Aquinas. Everything worth knowing about the natural world has already been written down.

This medieval view of science is separated from ours by two principal ideas. Firstly, it was religion, not science, which really concerned people in those days. The major task of medieval learning, at least in Europe, was to clarify our relation to God, not our relation to nature. For that reason science interested scholars only in so far as it illuminated theology; they did not value it for its practical applications which, in any case, were minimal. Secondly, society had not yet adopted an idea which we now take for granted, the idea of progress. Utopia for them was in the next world; the golden age of this world was in the past, not like ours in the future.

In the 17th century things began to change. A major prophet of the practical value of science and technology to the welfare of society appeared early in that century. Francis Bacon's grand design (*Novum organum*) was to 'restore and exalt the power and dominion of Man himself, of the human race, over the Universe'. For him knowledge was power, the power to improve life on Earth by useful inventions.

The next three centuries proved Bacon to be right, and by the end of the 19th century scientific research had laid the foundations of completely new industries. Knowledge, as Bacon forecast, had met with power, and the successful applications of science had firmly established the idea that we can improve the world by our own efforts, the idea of progress. Utopia was now in this world, not in the next. And so if we had consulted the man and woman in the street towards the end of the 19th century they would probably have seen science as an active means of progress. 'Of course scientific research is worth bothering about!', they would have said, 'it has brought us better health, travel and communications and a much wider variety of goods and entertainment than ordinary people have ever had before.' Indeed ordinary people had come to regard the improvement of their standard of living as a proper objective of government and they looked to science to make it happen. Although, like Bacon, they valued science mainly for its contributions to material progress, they were still interested in its new ideas. In those days science was more intelligible to the layman than it is now and there was a considerable enthusiasm for popular science.

Today, so recent surveys tell us, we can expect much the same answer to our question about science; but it is likely to be less enthusiastic. The modern litany no longer prays to the Lord for deliverance from lightning, tempest, fornication and the other deadly sins of mankind, but prays to the government for deliverance from pollution, over-population, nuclear war and the other deadly sins of science. Nevertheless, the surveys do show that most people still hold science in high regard and believe it to be essential to progress. Broadly speaking society values science more for new things than for new ideas. Indeed most people think of science as a clever box of tricks, *a modern cargo cult*. This is not only because applied science has been so successful, but also because the ideas of modern science are increasingly difficult to understand.

In this climate of opinion it is, of course, relatively easy to see the importance of applied research to society. It is, however, more difficult for the man and woman in the street to understand the vital contribution of basic or 'pure' science to the continued health of applied research. Indeed there is, I believe, only one really fireproof argument to justify basic research to a society which has learned to value science mainly for its utility. It is the simple argument that basic research is the seed-corn from which the practical benefits which we expect to reap from applied science must eventually grow. This may seem a rather obvious point to labour, but I have read official report after report on the support of basic science and have found that this argument is never articulated with the force and clarity which it deserves. Perhaps the reason is that the authors of those reports enlisted the help of economists, who on this topic are worse than useless, instead of consulting the historians of science.

An essential point which demands a knowledge of history is that although there is an all too obvious need for applied research to be relevant to the needs of society, the well-meaning, and often self-righteous, demand that all research should be obviously relevant to those needs is a serious threat to the long-term future of science. The history of science shows that society has been well served in the past by basic research which, although it may have been influenced by social needs, was not constrained by them; it was largely guided by the internal logic of science. Even the principal exponent of the social uses of science, Francis Bacon, was concerned to make this point when he wrote (*Novum organum*):

Nature to be commanded must be obeyed.

It would, of course, be nice if we could enlist public support for science on account of its cultural value. The arts have monopolised the popular definition of culture for years, perhaps we could steal some of their prestige for science. It shouldn't be impossible; after all the ideas of modern science are no more outrageous and lacking in popular appeal than many of the pictures and objects which we are called upon to admire as art, and personally I find the picture of the world presented by modern science more interesting and inspiring than most modern art. Furthermore, as Tom Wolfe has so wittily shown (*The Painted World*), the cultural values of the arts are not determined by what the public likes, or even by what the people with money like, they are determined largely by a small clique of art experts. Could scientists perhaps learn their secrets?

What is the cultural value of basic science? In trying to answer this question I am reminded that the 19th-century philosopher F.H. Bradley defined philosophy as the finding of bad reasons for something which one believes by instinct. It is the word culture which always makes me feel uncomfortable. It conjures up visions of people peering at incomprehensible pictures in an art gallery while they try to think of something original or polite to say, or of people dancing round a maypole waving coloured handkerchiefs. My latest dictionary isn't much help, it defines culture as 'the appreciation and understanding of literature, arts and music'; it says nothing about science. If we were to accept this definition then clearly the most significant contributions of basic science to culture would be, not ideas and understanding, but new processes and product, such as developments in printing, photography, sound recording, radio, television and so on, all of which have had an obvious effect on literature, arts and music and all of which owe a lot to basic science.

However, I suggest that we forget that narrow definition and take the meaning of culture to include not only the arts and customs of a society but also the complex of perspectives, values and ideas which underlie its world-view.

In medieval times, as I have said, the world-view of western culture was closely linked to its religion. People looked to religion to answer the great questions about life and for guidance as to what was right and what was wrong. The Renaissance, the Enlightenment and the Scientific Revolution destroyed the medieval world-view, and today the world-view of most of us is more closely linked to science than to religion. In medieval times it was almost true to say that is didn't matter what you said as long as it was religious; today it is almost true to say that it doesn't matter what you say so long as it is scientific!

People will always want to know what there is in the world and why. In earlier times when they asked what is the Sun and why it was put in the sky, they would have been told that it is a ball of fire and was put there by God to give us light. Nowadays they would be told that it is a ball of hot gas, fuelled by the conversion of hydrogen into helium, that it got there by condensation from a gaseous nebula and that the question of why it is there has no scientific meaning.

To try and answer our questions of what and how, but not of why, is broadly speaking, the principal cultural function of science. The importance of comparing our beliefs with what we actually observe in the world cannot be overestimated. If we fail to maintain that link between belief and reality, then there is no longer 'nature's truth', there is only 'your truth' and 'my truth', and society is in danger of losing the important distinction between fact and fiction and between science and magic. You have only to look at the persecution of witches in the 17th century or at the racial theories of the Nazis to understand why Francis Bacon said, 'God forbid that we should give out a dream of our imagination for a pattern of the world.'

The pattern of the world which is progressively revealed to us by science is not a simple catalogue of facts. It shows us new perspectives of the world seen in the light of our latest knowledge. It brings us radically new and dynamic ideas without which our understanding of ourselves and of the world around us would remain stagnant. Furthermore the actual doing of scientific research promotes values which our society needs.

First let us look at the perspectives. Modern science teaches us to see many questions in much broader perspectives than before. Nowadays scientific studies of the environment, acid rain, the effects of nuclear war and so on, must all be based on a view of the planet as a whole, on a view which is not only international but is global. That is a particularly valuable lesson at a time when so many of our most important problems are no longer national, but have become global.

Moving farther away from home, astronomy, geology and biology have transformed the perspective in which we see ourselves in time and space. They have shown us that we live on a minor planet of a minor star in a galaxy of billions of other stars, and that our own galaxy is only one of billions of other galaxies. And that is not all; they have also transformed our picture of ourselves in time. We now see ourselves in an evolutionary perspective in which everything, galaxies, stars, chemical elements and living things have evolved from a primeval fireball. In this picture the whole of human history is no more than the tick of the cosmic clock.

In the long run this new picture of our place in time and space is likely to have a greater cultural impact than the great voyages of exploration in the 16th and 17th centuries. Most of our current ideas about the meaning and purpose of life, our philosophies and religions, were formed with a very different picture of the Universe in mind, a picture in which the Earth and human life were more central, more significant, in the scheme of things. To be reminded, by the sheer scale of the Universe, that man is not the measure of all things is, I suggest, a good thing, especially in societies which are becoming increasingly irreligious.

Not surprisingly a common complaint about these new perspectives is that they rob the world of meaning; for example Steven Weinberg, writing about modern cosmology (*The First Three Minutes*), remarks that: 'The more the universe seems comprehensible, the more it seems pointless.' No doubt that is true if you expect to find meaning and purpose by looking through a telescope. To my mind these questions are mysteries, and what this enchanting new picture of the universe can teach us is, not that the world is pointless, but that our existing speculations about its meaning and purpose should be set in a much vaster frame.

Now what about the values of science? Science we are often told has nothing to do with values; it can tell us only what we *can* do and not what we *ought* to do. But we have only to look at the whole host of ethical problems raised by genetic engineering, nuclear power or by the many other consequences of scientific research to realise that, as the British monk Pelagius told us in the fifth century, it is often difficult, if not impossible, to separate ought from can. In that sense science has a great deal to do with values. It may not provide an adequate rationale for benefaction, but it certainly helps us to do good and to recognise evil. As William Blake wrote (*New Jerusalem*):

> He who would do good to another must do it in Minute Particulars. General Good is the plea of the scoundrel, hypocrite and flatterer, for Art and Science cannot exist but in organized Particulars.

I must also point out that scientific research is not just the accumulation of knowledge, but is an activity whose success depends upon respecting certain values, the most important of which Robert Merton (*Sociology of Science*) called 'organised scepticism', or in other words, a respect for the truth of fact. One of the greatest dangers to any society is that it should become credulous; the antidote to credulity is scepticism, a passion for the truth of fact, and that is something which is actively inspired and promoted by basic research. As Albert Schweitzer might have said, science teaches us reverence for truth, the truth of fact. Judging by the many cults of unreason, such as scientology, which flourish in our society today, by the fact that we have more astrologers than astronomers and that many of our students appear to believe in creationism, there is a need to promote the value of organised scepticism more actively.

But what about the undesirable effects of science on our values? If we listen to the prophets of the counter-culture as they emerge from the wilderness of industrial civilisation, we hear that science is a spiritual cul-de-sac and that, if we want to save our souls, we must back out of it. Science, so its critics say, has taught us to see the world as an impersonal machine which can best be understood by analysing it into component parts. This Mechanical Philosophy, so they say, has been damaging to society because many of our most pressing problems, particularly social problems, cannot be solved by the analytical methods of science and must be seen as a whole. Science is accused of dehumanising and disenchanting our world-view.

Throughout history these objections to the cultural influence of science have been voiced whenever science has been gaining ground; you have only to look at the works of Rousseau, Goethe, Blake or Wordsworth. Newton himself feared that too wide an application of his Mechanical Philosophy would disenchant the world by reducing the need for God. In our own day these same criticisms have been provoked by the enormous advances which science has made in the present century, culminating in the atomic bomb.

Their effect has been to make the public less friendly to science, to weaken the ideal of science as a vocation and, generally speaking, to put science on the defensive.

The only comments which I have space to offer are; firstly, that in my experience scientists are not insensitive to human and social problems, in fact they are to be found in the forefront of the various movements which support human rights, conservation and so on; secondly, that these objections are a warning that society must learn to control the practical applications of science more wisely; thirdly, that although science has certainly disenchanted the world by destroying superstition, it is busy re-enchanting it by giving us a picture of nature and of the Universe which is far more wonderful and inspiring than any of us could ever have imagined; and lastly that these criticisms are more appropriate to the classical science of the 19th century than to modern science which is significantly less mechanical and more organic in character.

Now what about the ideas of science? Ever since modern science began, its ideas have been remaking the world. Indeed they were so influential in the 16th and 17th centuries that the historian Herbert Butterfield (*Origins of Modern Science*) writes of the scientific revolution that,

> . . . it outshines everything since the rise of Christianity and reduces the Renaissance and Reformation to mere episodes, mere internal displacements, within the system of medieval Christendom.

The Mechanical Philosophy of Descartes and Newton certainly penetrated into every branch of the culture of the 16th and 17th centuries; and the ideas of Darwin and Wallace on evolution, of Freud on the unconscious, and of Einstein on relativity have all been profoundly influential in the 19th and 20th centuries. The same will, I feel sure, prove to be true of the many other new and strange ideas of 20th-century physics.

In the present century our exploration of the world with new and more powerful instruments has exposed the limitations of classical science. Much to our surprise we have learned that the behaviour of things which are very large, very small or moving very fast is exceedingly strange and cannot be understood or predicted by the common-sense methods of classical science. Indeed to construct theories which predict this behaviour we have had to revise many of the ideas which, up till now, we had regarded as the very pillars of scientific thought.

As a consequence many of the sacred cows of classical science have sickened and died. For example, we have been obliged to recognise that our classical ideas about the role of the observer in physics are largely illusory. We now realise that all our descriptions of things are not of their intrinsic properties but are metaphors which describe their behaviour when observed in a certain way; this leads to the sort of conceptual paradox which we meet in the modern picture of light, in which light is pictured both as a particle and as a wave. In many ways this problem is analogous to that encountered by Christian theologians in their attempts to describe another mystery, the threefold nature of the Trinity.

As another example we have found that our classical ideas about causality do not apply to atomic events and that in predicting individual events we must exchange certainty for probability; even the sacred cows called identity and locality look a bit sick. It rather looks as though physicists will have to give up breeding sacred cows and only breed an animal which is more useful but has a shorter life, the hard working hypothesis. Modern science has not given up the ambition of explaining the world, what it has done is to change what it means by the word explanation.

Many of these new ideas are so abstract and strange that it is difficult to foresee what their cultural effect will be. The classical science of the 19th century, the so-called Mechanical Philosophy, established an image of science which was ruthlessly analytical, mechanical and, generally speaking, lacking in human qualities; it was seen to be supremely self-confident and was widely regarded as the only source of authentic knowledge about the world. By comparison modern science is far less mechanical, more concerned with interactions and form and, generally speaking, more human; it is also more aware of the nature and limitations of scientific knowledge.

A growing appreciation of these new qualities of modern science and of the fact that there is a significant difference between the ideas of science and their practical applications, has already blunted the attacks of the counter-culture on the cultural effects of those ideas. Indeed diatribes against the ideas of science are being replaced by efforts to find common ground; for example by those many recent books which draw rather foggy parallels between modern physical thought and mysticism.

No doubt the acceptance of uncertainty and metaphor by modern physics tends to blur the essential differences between real science and the cults of unreason; nevertheless let us hope that, in the long run, these new qualities of modern science will make it easier to forge links between science and the other branches of our culture, such as art, philosophy and religion. We must forge those links if we are ever to repair the disastrous effects of the modern fragmentation of knowledge and its separation from faith, and so to arrive at a more coherent world-view.

If the 19th century taught us that knowledge is power, the 20th century has taught us that our ability to produce new knowledge greatly exceeds our ability to use the power which it brings wisely. Obviously if we are to get more of what we want and less of what we don't want from science, we must develop wiser ways of choosing how we apply it – better forecasting, assessment and so on.

But there is something else. As I see it, our best hope of living happily is to try to gain a better understanding of ourselves and of the profoundly mysterious world in which we find ourselves. Modern science can tell us more about the world than any society has ever known before and if we are to make wiser use of that knowledge, we must learn to treat science as an integral and valuable part of our culture and not simply as an agent of material progress. We must accept that the most influential ideas are not necessarily the most practical.

As Jacob Bronowski pointed out (*Science and Human Values*),

> The body of technical science burdens us because we are trying to use the body without the spirit.

We must recognise that science, particularly basic science, does have a distinctive and valuable spirit and, as I have tried to show all too briefly, it is a spirit which most modern societies, ruled as they are by the values of economics, would do well to encourage.

R. Hanbury Brown

1

Astronomy in ancient and medieval China

JOSEPH NEEDHAM

It is at present most appropriate that this volume on *Cosmic Perspectives* is aimed at stimulating discussions of the nature of scientific progress, and of its social conditions and consequences, and I should like to congratulate the editors for their dedicated work in this regard. Forty years after the use of nuclear weapons against the cities of Hiroshima and Nagasaki, and in the light of the massive problems of social development and welfare which now loom before us, such discussions are clearly of the utmost importance. Since I myself am now primarily engaged in the comparative history of traditional Chinese science, technology and medicine, my contribution to this volume will be devoted to the history of astronomy in China.

Chinese astronomy differed from that of the Western world in two important respects: first it was polar and equatorial rather than planetary and ecliptic, and secondly it was an activity of bureaucratic States rather than that of priests or independent scholars. Both qualities had advantages and disadvantages. The first, the polar–equatorial situation, led to the mechanization of celestial models far earlier than in the West, but it deferred the recognition of equinoctial precession till later. Similarly, the bureaucratic situation ensured the recording of remarkable sets of celestial observations, but on the other hand it probably discouraged causal speculation, especially since the Chinese did not have Euclidean deductive geometry.

I think most people are well aware that from our present point of view nothing much was happening in China until about 1500 B.C. when the Shang period started, and after that during the Chou period there remain various records. But we have the fullest details from the Chhin and Han onwards, from about the 3rd century B.C. through the many centuries during which the dynastic histories were written, recording an enormous amount of astronomical information. Many people think that it is necessary to rely upon manuscript sources, as for example our friends the Arabists have to do in great part, but fortunately this is not the case in China because broadly speaking one can say that everything there is either printed or lost. Besides this, there probably are archives containing important astronomical data, which have not yet been opened in China itself and also in countries like Korea, and I think that some considerable discoveries may be expected in the future on the basis of such archives.

Now as regards the polar and equatorial character, it is clear that Chinese astronomy was of this kind from the beginning. The calendrical problem was of course the simultaneous observation of the stars and the Sun, and presumably there are only two possible methods for ascertaining this relation, methods which people have called contiguity and opposability. The method of contiguity was, we know, that of the ancient Egyptians and the Greeks; it involved the observations of heliacal risings and settings, i.e. rising and settings just before sunrise and just after sunset. We all remember the famous heliacal rising of Sirius in ancient Egypt. This kind of observation did not require knowledge of the pole, meridian or celestial equator, nor any system of horary measurement. It naturally led to the recognition of the zodiacal or ecliptic constellations, and of stars appearing and disappearing simultaneously with them nearer or farther away from the ecliptic, the paranatellons as they used to be called. But the opposability method was that which was adopted by the ancient Chinese; they never paid much attention, so far as any of the records show, to heliacal risings and settings, but rather to the pole star (*pei chi*) and the circumpolar stars which never rise and never set. Their astronomical system was associated with the concept of the meridian, which would arise naturally out of the use of the gnomon, and they systematically determined the culminations and lower transits of the circumpolar stars.

I think one could really say that the celestial pole was the fundamental basis of Chinese astronomy. It was connected also with the microcosmic–macrocosmic type of thinking, because the pole corresponded to the emperor on Earth, around whom the vast system of the bureaucratic agrarian state revolved naturally and spontaneously. I mentioned the gnomon (*piao*) just now, and clearly the meridian concept would arise very easily from this upright stick, because if you looked south you could measure the noon shadow, and if you looked north during the night you could measure the times at which the various circumpolars made their upper and lower transits. We have this in so many words in Chinese texts. Moreover, just as the influence of the Son of Heaven on Earth radiated in all directions, so the hour-circles radiated from the pole. During the 1st millennium B.C. the Chinese built up a complete system of equatorial divisions defined by the points at which the hour-circles transected the equator (*the chhih tao*); and these divisions contained the 28 'lunar mansions' (*hsiu*), like segments of an orange filling up the celestial sphere, bounded by hour-circles and named

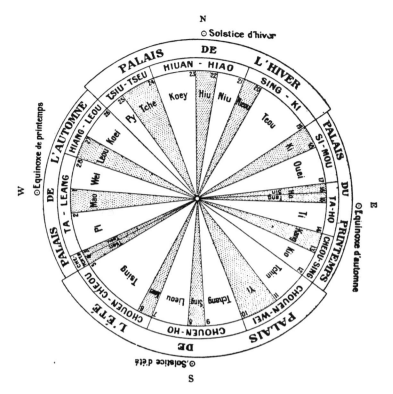

Fig. 1.1. Diagram of a projection of the lunar mansions (*hsiu*) on the equator, done for the 24th century B.C. The wide variation in their extensions may be noted.

from the constellations providing the 'determinative stars' of boundary markers (*chii hsing*). The classical diagram of de Saussure is given in Fig. 1.1.

Having once established the boundaries of the *hsiu* by means of the characteristic asterisms scattered around the equator, and their determinative stars, the Chinese were in a position to know their exact locations, even when invisible below the horizon, simply by observing the meridian passages of the circumpolar stars which they could always see. Since they knew where the equator asterisms were at all times, they were able to solve the problem of the sidereal solar positions, because the sidereal position of the full Moon was in opposition to the invisible position of the Sun. In the same way we find growing up quite early in the late Chou, the Warring States period and from the beginning of the Chhin and Han, from the 3rd century B.C. onwards, a rather full and complete recognition of the great celestial circles. This is known as the Hun Thien cosmological system, but it was not really a cosmology as often so called, it was rather the recognition of these circles. Naturally it accompanied the development of observational and demonstrational armillary spheres.

As regards the origin of the system of the *hsiu* or lunar mansions, this is a very difficult question because we get other systems of the same kind in other civilizations: particularly the *nakshatra* in India and the *manāzil* in the Arabic world. The *manāzil* do not compete of course but indianists and sinologists have long been disputing about which is the older of the two, the Indian or the Chinese. I cannot today go into the argument on both sides, but one can say that only nine of the twenty-eight *hsiu* determinatives are identical with the corresponding *yogatārās* or 'junction stars' of the Indians, while a further eleven share the same constellation but not the same determinative star. Only eight of the determinative stars and *yogatārās*, however, are in quite different constellations, and of these two are Vega and Altair. On the Chinese side it is possible to say that the *nakshatra* do not show so clearly the coupling arrangements whereby *hsiu* of greater or lesser equatorial breadth stand opposite each other. Indian astronomy moreover, which was far more influenced by Greek astronomy than the Chinese was, does not show that keying of the *hsiu* and the circumpolar stars which is so important in China, in fact the essence of the Chinese system. Besides, the distribution of the *nakshatra* asterisms is much more scattered than that of those of the *hsiu*, following even less closely the position of the equator in the 3rd millennium B.C.

On the other hand, as regards the documentary evidence, the Indians have very little to yield. There seems to be not much doubt that in the hymns of the Rig Veda, which correspond with the Shang oracle-bones and come down from about the 14th century B.C., two *nakshatra* make their appearance. From that time onwards gradually the system is built up. It becomes complete in the *Atharva Veda*, for instance, and in the black *Yajur Veda* (all three recensions). This may mean that the system was fairly organized in India by about 800 B.C. But in China again there is much the same situation because that ancient calendar called the *Yüeh Ling* may come down from as far back as 850 B.C. and it mentions twenty-three out of the twenty-eight *hsiu*. We cannot pursue the long argument but nevertheless the problem is a very interesting one and it is not yet solved. I myself have always wanted to believe that the original circle of lunar mansions round the equator was Babylonian. The only difficulty is that it may be rather hard to find anything in Babylonian astronomy which could really have given rise to the *hsiu* and *nakshatra* systems.

May I now come to the question of celestial coordinates. In China we have a remarkable text called the *Hsing Ching* or 'Star Manual', the date of which is very unsure, though undoubtedly ancient. Some of its data appeared also in much later works such as the 8th century A.D. *Khai-Yuan Chan Ching*. The *Hsing Ching* is certainly Han, but it may be a little earlier, and it records the positional measurements of many stars made by three astronomers of the 4th century, Shih Shen, Kan Tê and Wu Hsien. Some of them are *hsiu* or lunar mansion stars, and some of them are other stars all over the heavens grouped in asterisms. The epoch of the observations is not at all certain, and calculations have shown varying dates. None of the observations could be earlier than about 350 B.C. but some are a good deal later, so the catalogue as a whole may or may not be pre-Hipparchan (134 B.C.). Some of the measurements seem to have been made about A.D. 130 nearer Ptolemy's time. The text gives the name of the asterism, the number of stars it contains, its position with respect to neighbouring asterisms, and measurements in degrees (on the $365\frac{1}{4}°$ basis of course) for the principal stars in the group; i.e. the hour-angle of the principal star measured along the equator from the first point of the *hsiu* in which it lies, and the north polar distance of the star. The text says, for example, that such and such a star is 2° forward from the beginning of

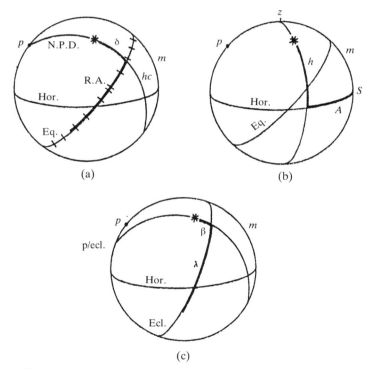

Fig. 1.2. The three systems of celestial coordinates; (*a*) the equatorial Chinese, and modern system; (*b*) the Arabic altazimuth system; (*c*) the Greek ecliptic system.

the *hsiu Hsin*, and also that its distance from the north pole is 103°. Obviously the first corresponds to our modern right ascension, and the second gives the complement of our modern declination.

This is really quite interesting because I think it is well known (Fig. 1.2) that the Greek coordinates were essentially ecliptic, positions being measured along the ecliptic and towards the pole of the ecliptic. It is equally well known that the Arabic system made use of the horizon taking the azimuth and altitude. This has the great disadvantage that it applies only to particular individual points on the Earth's surface. In China we never get the horizon system at all, or only perhaps extremely late under Arabic influence. On the other hand, the Greek system does make its appearance in the Thang period, when some of the Indian astronomers were working in China, for example a text may say that a star is so many degrees north or south of the ecliptic; but this is a late thing found only in the 8th, 9th and 10th centuries A.D. All civilizations have used one or other of these three different types of star coordinates and our modern ones are clearly Chinese.

Now as for the constellations and the naming of them, it is rather an interesting thing that there is no overlap at all with the nomenclature in other civilizations. There may be a good deal of connection between Greek and

Fig. 1.3. The Moon passing through a constellation; from a Szechuanese moulded brick of Han date (from Wen Yu). The Moon disk shows a toad under a tree (a legendary consequence of the lunar crater markings, etc.) and is borne along by a feathered and winged genius. The constellation is drawn in the usual ball-and-link convention.

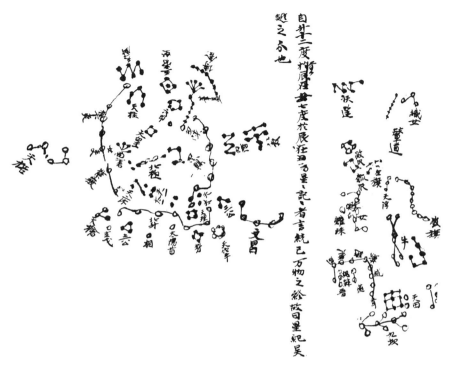

Fig. 1.4. The Tunhuang MS. star-map of A.D. 940. To the left, a polar
projection showing the Purple Palace and the Great Bear below it.
To the right, on a 'Mercator' projection, an hour-angle segment
from 12° in Tou *hsiu* to 7° in Nü *hsiu*, including constellations in
Sagittarius and Capricornus. The stars are drawn in three colours,
white, black and yellow, to correspond with the three ancient schools
of positional astronomers (those of Shih Shen, Kan Tê and Wu
Hsien).

Indian names but practically nothing in China corresponds. From ancient
times it was customary to represent constellations in a ball-and-link style, as
we see on an inscribed brick from the Han period (1st century B.C. to 1st
century A.D.), Fig. 1.3. Fig. 1.4 shows an early star-map from the Chinese
culture area datable at about A.D. 940. This manuscript must be one of the
oldest star-charts extant from any civilization. It gives a picture of the north
polar region, the 'Purple Forbidden Palace', as it was called, in which you can
see down at the bottom the shape of the Great Bear similar to the way
everybody else saw it, but nothing else is the same. A European constellation
can appear in several different asterisms on the Chinese planisphere; for
example Hydra comprises the three *hsiu* Chang, Hsiang and Liu, together

with eight other star groups having no similarity of symbolism. And this is the case all along the line. There is nothing corresponding to Delphinus and Cancer and so on, the only overlaps are the Great Bear, or Northern Dipper (Pei Tou as we call it in Chinese), and of course naturally Orion (Shen) and the Pleiades (Mao), but apart from those there is simply no correspondence. What is rather interesting is that the agrarian bureaucratic nature of Chinese civilization led to a multitude of star and constellation names in which the hierarchy of earthly officials had their heavenly counterparts.

Next, Fig. 1.5 shows what one can only describe as a star-map on a 'Mercator' projection, with the upright bands representing the *hsiu*, the lunar mansions of different widths or stretches, and then the equator running along the middle the ecliptic also shown over half the heavens, and all the stars

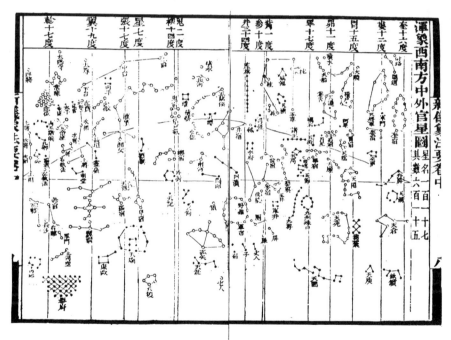

Fig. 1.5. Star-chart from Su Sung's *Hsin I Hsiang Fa Yao* (A.D. 1094), showing fourteen of the twenty-eight *hsiu* (lunar mansions), with many of the Chinese constellations contained in them. The equator is marked by the central horizontal line, the ecliptic arches upwards above it. The legend on the right-hand side reads: Map of the asterisms north and south of the equator in the south-west part of the heavens as shown on our celestial globe; 615 stars in 117 constellations. The *hsiu*, reading from right to left are: Khuei, Lou, Wei, Mao, Pi Tshui, Shen (Orion), Ching, Kuei, Liu Hsing, Chang, I and Chen. The very unequal equatorial extensions are well seen.

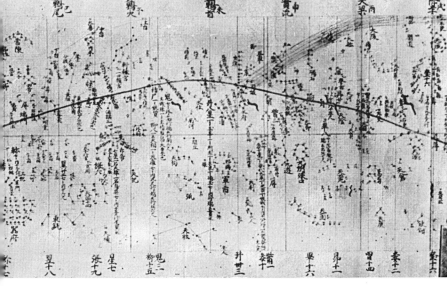

Fig. 1.6. The same hemisphere as in Fig. 1.5, with the equator again horizontal and centre, and the *hsiu* divisions and constellations shown as before, on a 'Mercator' projection: from an MS. star-map Koshigettshin-zu formerly preserved in Japan (Imoto Susumu). Many similarities with the preceding figure will be apparent, but the drawing has been done on squared paper for greater accuracy. Perhaps the original was more likely Sung than Thang.

placed in their approximately right positions. This was the map of the stars which was made for the celestian globe set up by Su Sung in the astronomical clock tower at Khaifêng, in A.D. 1088 some 500 years before Gerard Mercator. I shall want to say a word more about this instrument in a moment. A similar star-chart in Japan has been described by Dr Imoto Susumu (Fig. 1.5 gives the same hemisphere as Fig. 1.6). But it has the additional interest of being drawn on squared paper with the aim of greater accuracy. Though the manuscript is of a later date, Imoto Susumu traces its origin back to the time of I-Hsing in the 8th century A.D., since the R.A. values mostly though not entirely, agree with those given by him. Su Sung may have drawn from the same source. Finally, we show the famous planisphere of Suchow (A.D. 1193) in Fig. 1.7.

Coming back to the question of instrumentation, the earliest measuring device was undoubtedly the gnomon. We have many references to that in ancient books, such as the *Tso Chuan* and the *Chou Pei Suan Ching*. It must have been used from the Shang period, *c.* 1500 B.C. onwards. The *Chou Pei*

Fig. 1.7. The Suchow planisphere of A.D. 1193 (Rufus and Tien). Note the excentric ecliptic and the curving course of the Milky Way (*thien ho*, the river of heaven). The map with its explanatory text was prepared by the geographer and imperial tutor Huang Shang, and committed to stone by Wang Chih-Yuan in A.D. 1247.

Suan Ching (Arithmetical Classic of the Gnomon and the Circular Paths of Heaven) and the *Chou Li* (Institutes of the Chou Dynasty) are both probably early Han texts but based on the usages of the centuries preceding. In many ways, the gnomon was a much used instrument in Chinese culture. An interesting new light has recently been thrown on this, relating especially to the meridian line of observation posts set up about A.D. 725 under the

Buddhist monk I-Hsing, the greatest astronomer and mathematician of his age. Now this was a chain of stations reaching from Indo-China right up to Siberia, and the seasonal observations of noon Sun shadows with the gnomon were done at a dozen places along that line. Measuring just over 2500 km in length, this meridian survey must be one of the greatest efforts of organized scientific research in any medieval civilization.

The gnomon reached its climax in Chinese culture with the giant instrument set up by Kuo Shou-Ching about A.D. 1276 (Fig. 1.8). This shows the long scale for the measurement of the Sun's shadow thrown by the 12 m gnomon, and the star observation platform is to be seen at the top of the tower. One of the chambers there housed a clepsydra, perhaps a hydro-mechanical clock; the other probably an armillary sphere. This wonderful piece of giant astronomical equipment still exists at a place called Kao-Chheng not far from Lo-yang (in *Honan*) near the geographical centre of China, and it

Fig. 1.8. The Tower of Chou Kung for the measurement of the Sun's solstitial shadow lengths at Kao-Chheng (formerly Yang-Chheng), some 80 km south-east of Lo-yang, and for many centuries the site of China's central astronomical observatory. The present structure is a Ming renovation of the instrument built by Kuo Shou-Ching about A.D. 1276. The 12 m gnomon stood up in the slot, and its shadow was measured along the stone scale projecting on the left, with special arrangements to secure a sharp edge reading. One of the rooms on the platform housed a clepsydra (perhaps a hydro-mechanical clock), and the other probably an observational armillary sphere.

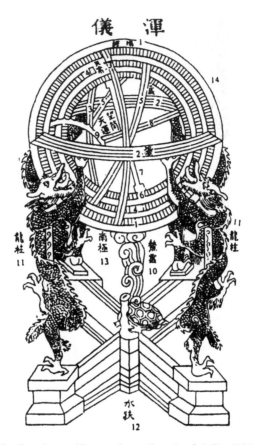

Fig. 1.9. Su Sung's armillary sphere (*hun i*) of A.D. 1088, described in the
Hsin I Hsiang Fa Yao redrawn from the text and labelled by
Maspero.

has been repaired in very recent times, in fact even during the Cultural
Revolution. During the war it was used for target practice by the Japanese,
and the top of one of the side chambers was knocked off, but no other
damage was done, and it has now been completely repaired and restored. It
must be regarded as a notable monument even though no doubt it was rebuilt
under the Ming. The 13th century A.D. data obtained with this gnomon still
exist also, and are highly creditable for that time, especially as a sophisticated
device, the 'shadow definer', was used to focus the image of the cross-bar and
avoid the difficulty of the penumbra.

The development of armillary rings and spheres is an even more intriguing
question. It is fairly certain that the most primitive form of armillary was a
simple single ring with some kind of fiducial line or sights which could be set

Fig. 1.10. The equatorial armillary sphere of Kuo Shou-Ching (*c*. A.D. 1276),
now preserved in the grounds of the Purple Mountain Observatory,
Nanking.

up in the meridian or equatorial plane. Measurement one way gave the north
polar distance (N.P.D.) or declination, measurement the other way gave the
position in the *hsiu*, i.e., the right ascension. No doubt that was all that Shih
Shen and Kan Tê had at their disposal, and there is some evidence that it
came down to about 100 B.C. because it may be that Lo-Hung and Hsien-Yü
Wang Jen had nothing else. But then things happened rather fast. Kêng
Shou-Chhang introduced the first permanently fixed equatorial ring in 52
B.C. and the ecliptic ring was added by Fu An and Chia Khuei in A.D. 84;
while with Chang Hêng's apparatus in A.D. 125 the sphere was complete
with horizon and meridian rings. It is rather remarkable that this rapid
evolution should have come about historically parallel with Greek times and
just before the life of Ptolemy himself.

Spheres of all kinds continued to be made without much change for many
centuries. The one in Fig. 1.9 is the very famous instrument constructed by
Su Sung in A.D. 1088 or so for the astronomical clock tower at Khaifêng
which I mentioned before. In the Thang period, about A.D. 630, Li Shun-
Fêng made the radical innovation of building not two nests of concentric
rings but three, so it got rather over-complicated, but the design that Su Sung
used is one which has many similarities with the equatorial spheres of Tycho
Brahe in 16th century A.D. Europe. Fig. 1.10 shows another sphere, that of
Kuo Shou-Ching (A.D. 1276) which is to be found at the Purple Mountain

Fig. 1.11. Kuo Shou-Ching's 'equatorial torquetem' (*chien i*, simplified instrument), precursor of all telescope equatorial mountings, in the grounds of the Purple Mountain Observatory, Nanking. Like the armillary sphere in Fig. 1.10, this instrument may be one of the identical replicas cast by Huang-Fu Chung-Ho in A.D. 1437 (orig. photo 1958).

Fig. 1.12. Detail view of part of the instrument shown in Fig. 1.11 (orig. photo 1958). The bronze mobile declination split-ring or meridian double circle carrying the sighting-tube is well seen. Below, the fixed diurnal circle, and the mobile equatorial circle, with movable radial pointers probably used to demarcate the boundaries of *hsiu*.

Observatory near Nanking today. This was the same astronomer who made the giant gnomon device. But his major achievement was his equatorial torquetum or 'simplified instrument' (*chien i*), which did away with the unnecessary parts of the armillary sphere and achieved what was essentially an equatorial mounting of the sighting tube. One can still see it (Figs. 1.11 and 1.12) on the top of the Purple Mountain at Nanking, and I am glad to report that it is in perfect order and very carefully preserved.

This brings me to the question of powered celestial models which I men-

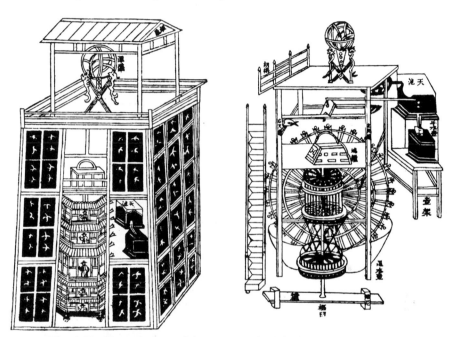

Fig. 1.13. General views of the astronomical clock tower built at Khaifêng by Su Sung and his collaborators in 1090. On the top platform, some 11 m above the ground, there was a mechanized armillary sphere for observations; in a chamber on the first floor a mechanized celestial globe was installed, and below, in the form of the hydro-mechanical clockwork, numerous jacks appeared at the openings of a pagoda façade, constituting a time-annunciator. The tower and its machinery are fully described in Su Sung's *Hsin I Hsiang Fa Yao*. (Left) External appearance, with a panel removed to show the constant-level tank. (Right) Internal structure. In front, the jack-wheels and vertical shaft, behind this the driving wheel. On the right the constant-level tank delivering the water to the driving wheel, scoops or buckets. Above the driving wheel one can see a few traces of the escapement mechanism (*cf.* Figs. 1.17, 1.18). On the left, the staircase, on the top platform, the armillary sphere.

tioned earlier, and that certainly was an extremely interesting development unexampled in any other culture. Twice already I have referred to the astronomical clock tower of the late 11th century A.D. described in the *Hsin I Hsiang Fa Yao* (*New Design for a Mechanized Armillary Sphere and Celestial Globe*), a book presented to the throne in A.D. 1092. These cosmic models, in fact demonstrational armillary spheres and celestial globes, were rotated by water power using a constant-level tank and a driving wheel with buckets. One might call it a mill wheel but it was retained and guided all the time by an escapement so we use the term hydro-mechanical clockwork, and speak of the hydro-mechanical linkwork escapement. The background of the constant-level tank lay in clepsydra technology, for in the 6th century A.D. the older polyvascular trains of compensating tanks had been superseded by arrangements of overflow tanks to produce a perfectly steady flow.

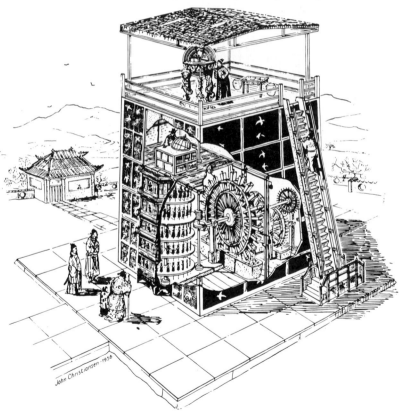

Fig. 1.14. Pictorial reconstruction of the Khaifêng clock tower (J. Christiansen). Besides the components already mentioned, the norias which wound up the water back into the tanks are here glimpsed behind the driving wheel.

The escapement was indeed a great invention; it was certainly in use by the time of I-Hsing and Liang Ling Tsan at the beginning of the 8th century A.D., and the only problem is whether it may go back much further. It is still uncertain whether Chang Hêng in the 2nd century A.D. had already got this or not. Thus using the constant-level tank, and the driving wheel retained by the linkwork escapement, one had a real time measuring machine, and of course it may also be considered the first of all clock drives.

Fig. 1.13 shows the appearance of the clock tower, with the tanks and the driving wheel laid open on the right; and you can see the celestial globe on the first floor, and on the roof the armillary sphere. I called it demonstrational just now, but that is not quite fair because it was certainly an observational one too, and we have speculated that one of the reasons for introducing the

Fig. 1.15. Model of the Khaifêng clock tower in the Science Museum at South Kensington (J.H. Combridge). Here it is seen from the back and right.

clock drive was so as to be able to turn round the 20-odd tonnes of bronze in time to make an observation just before dawn. But the globe rotating in the intermediate storey would have been for demonstration in times of clouds and storm when the heavens could not be observed directly. Figs. 1.14 and 1.15 show reconstructions of what the clock tower looked like, and the mechanism follows in Fig. 1.16. Here one can see how the drive on the right hand side at the bottom rotated the column bearing much jack work, and a celestial globe on a level drive at the top, and how after some time the long shaft was replaced by a chain drive, in fact mark 2 and mark 3 chain drives, getting progressively shorter and so better designed. Finally the escapement mechanism is seen in Figs. 1.17 and 1.18; this is the device that is certainly of the 8th century A.D. and possibly of the 2nd century A.D.

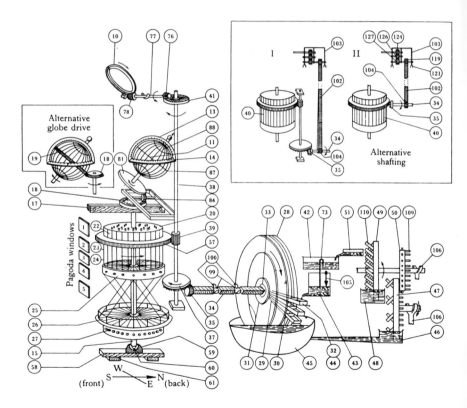

Fig. 1.16. Diagram of the power and transmission machinery of the Khaifêng clock tower (J. Needham, L. Wang and D. de S. Price). Norias and tanks on the right, driving wheel central, then the main vertical shaft (afterwards replaced by successively shorter chain drives) operating jack-wheels, globe and armillary.

Fig. 1.17. Working model of the hydro-mechanical escapement of the Khaifêng clock tower in the Science Museum at South Kensington (J.H. Combridge).

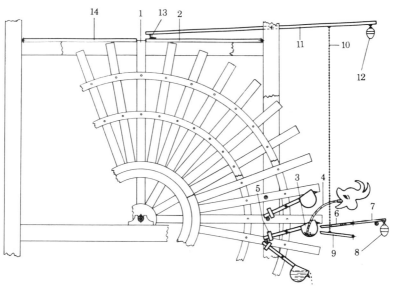

Fig. 1.18. Diagram to illustrate the functioning of the hydro-mechanical linkwork escapement (J.H. Combridge). Each scoop bucket on the perimeter of the driving wheel is individually balanced; when it is full it descends, depressing a counter-weighted trip-level, and operates a chain-and-link connection which opens a gate at the top of the wheel and allows the next scoop bucket to come into place. One 'tick' took place about every 24 s.

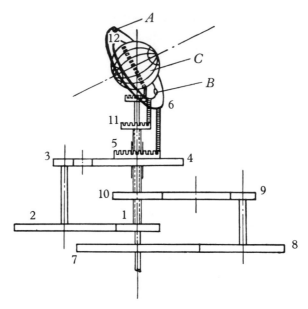

Fig. 1.19. Reconstruction of the orrery movement of Chinese hydro-mechanical astronomical clocks (Liu Hsien-Chou). Only the Sun and
Moon model movements are shown as well as that for the celestial
globe, in a design which would have needed both concentric
shafting and gear wheels with odd numbers of teeth. Remaining
textual descriptions authorize it, however, for the instruments of
I-Hsing (A.D. 725), Chang Ssu-Hsün (A.D. 979) and Wang Fu
(A.D. 1124).

Now these powered models illustrate, I think, the advantage which the
Chinese found in having a polar equatorial system. After all, the celestial
latitude and longitude grid is a purely conceptional network thrown over the
heavens, and along its lines nothing ever actually moves, but things of course
do move on the circles parallel with the equator. Finally, Fig. 1.19 shows a
reconstruction by Dr Liu Hsien-Chou of I-Hsing's orrery arrangement for
Sun and Moon models as deduced from the textual descriptions of his early
8th century A.D. astronomical clock.

Let me return lastly to the theme of bureaucratism that I opened with,
because it is really a very interesting feature, a characteristic of Chinese
civilization which made it quite different from all others in those ancient
times. One might say that the majority of the observers who thought and
calculated and wrote about astronomical problems through 2000 years were
in State service. They were organized in a special department of government,
the Astronomical Bureau or Directorate, which went by various different

names. The most ancient title of the Director was Thai Shih Ling, and although we have always been very conscious of his astrological function we feel that the true astronomical and calendrical element in the work of his department was amply sufficient to warrant the translation 'Astronomer-Royal'. He did of course have to keep his star-clerks on the watch, every moment of every night, for any unusual developments in the heavens. Celestial phenomena like novae, supernovae, eclipses, comets, meteor streams, sun-spots and all such things were regularly reported to the Imperial Court, because although individual genethliacal astrology came only very late to China, perhaps in the Ming, the general belief that 'comets do foretell the death of princes' was a very old Chinese idea. The origin of the Bureau of Astronomy was thus perhaps twofold; the importance of keeping the calendar in order was a very important task, but the watch on the heavens for celestial events was also a strong motive. It is interesting that inauspicious happenings were generally regarded as *Chhien Kao*, or reprimands from heaven, and the emperor or some high official, very often the emperor himself, took the guilt upon him, prayed, fasted and promised to amend. Omens were regarded really as signs of bad government, and there would be trouble if things were not put in order; such was the astrological functions of the Bureau over the ages.

One remarkable fact not generally known is that in some periods it was customary to have two observatories at the capital both furnished with armillary spheres, clepsydras and all manner of necessary apparatus. For example, Phêng Chhêng tells us about this in the Northern Sung. The Astronomical Department called the Thien Wên Yuan was located within the walls of the Imperial Palace itself, while the other one, the Directorate of Astronomy and Calendar, the Ssu Thien Chien, presided over by the Rhai Shih Ling himself, was outside the walls. The data from the two observatories especially concerning unusual phenomena, were supposed to be compared each morning and then presented jointly so as to avoid false or mistaken reports.

I have mentioned the grave political significance of celestial events. It is therefore rather interesting to find exhortations to security-mindedness addressed century after century to the astronomical officials. For example, the *Chiu Thang Shu* (*Old History of the Thang Dynasty*) tells us that in A.D. 840 in the twelfth month of the 5th year of the Khai-Chhêng (reign-period) an imperial edict was issued ordering that the observers in the Imperial Observatory should keep their business absolutely secret. 'If we hear,' it said, 'of any intercourse between the astronomical officials or their subordinates, and officials of any other government departments, or miscellaneous common people, it will be regarded as a violation of security regulations which should be strictly adhered to. From now onwards, therefore, the astronomical officials are on no account to mix with civil servants and common people in general. Let the Censorate see to it.' All one can say about that is that there is nothing new about Los Alamos or Harwell, but whether or not the greatest scientific achievements happen under such conditions is another question. And I suppose that even a Galileo and a Priestley had their difficulties with the powers that were.

Something of what the observatories did has come down to us in extant texts. Records of eclipses start with the Shang oracle-bone material from 1361 B.C. onwards, and records on novae from 1300 B.C. (Fig. 1.20).

The oracle-bones are of course difficult to interpret, but nevertheless they provide evidence of great importance. As for supernovae, there was the famous one of A.D. 1054, the Crab supernova. Then there is the abundance of documentation on comets. Fig. 1.21, shows a manuscript drawing of a comet passing between two *hsiu* constellations on the night of 28 October A.D. 1664; it comes from the Korean archives. For comets in general we have Chinese records from 613 B.C. onwards. Finally sun-spots were being sedulously recorded from 28 B.C. onwards, a fact which I think would have been a great surprise to Galileo and Christopher Scheiner if they had ever been aware of it.

My last remarks must concern cosmologies. The Kai Thien cosmology, with a domical arrangement of the Earth and the heavens, was rather Babylonian, obviously very primitive (Fig. 1.22), and did not play much part

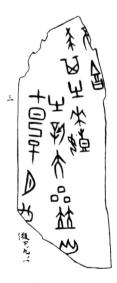

Fig. 1.20. The oldest record of a nova; inscription on an oracle-bone dating from about 1300 B.C.

Fig. 1.21. A comet of A.D. 1664, from the Korean astronomical archives (Rufus).

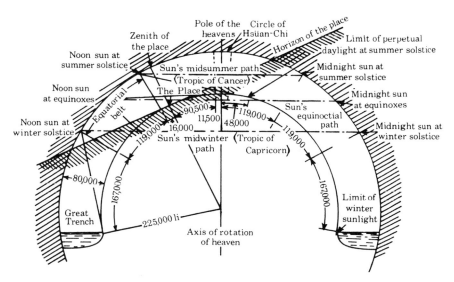

Fig. 1.22. The Kai Thien cosmology (Chatley).

after the Chhin and Han in China. At this time there was perfected the Hun Thien cosmology which was really the recognition of the great celestial circles. Finally there was the Hsüan Yeh cosmology, which is of greater interest because it maintained that stars were lights of uncertain origin, floating in infinite empty space, and that the planets and moving stars were carried round in this dark space by some kind of wind. Since Euclidean deductive geometry was not available, no Ptolemaic geometrical model was devised and all through the ages calendrical calculations were done by algebraic methods, thought about the actual mechanics and geometry of the solar system being laid aside. Perhaps that preference was again a Babylonian characteristic. In any case the Hsuan Yeh theory was rather modern and long before its time. One could say that if the Chinese had no Euclidean geometry they had no crystalline celestial spheres either, so they did not have to break out of them at the time of the Renaissance; and it may even be that a knowledge of the Chinese conceptions helped Giordano Bruno, William Gilbert and Francis Godwin to take this great step themselves.

To sum it all up, Chinese astronomy cannot be neglected by anyone who seeks for an oecumenical account of the development of human knowledge of the starry firmament and our own place within it. It is all the more important on account of its extreme originality, influenced indeed a little by Babylonia and India, but unlike the latter culture highly independent of those Greek and Hellenistic discoveries which had so wide-ranging an influence everywhere west of Turkestan.

2

Indian astronomy: an historical perspective

B.V. SUBBARAYAPPA

The general Indian ethos, like that of the other ancient culture-areas, was one of an integrated man–spirit–cosmos view, a wide and comprehensive view of Nature in which the *Homo sapiens*, or Man, the thinker, occupied a distinct place. For quite some time, this triple-strand cosmic view was mixed up with several facets of mythological elements. However, the Cosmic Man (*Puruṣa*), the Cosmic Light (*Viśvajyoti*) and the Cosmic Law (*Ṛta*) were perceived by Vedic seers (*Ṛsis*), discarding the mythological shackles as they endeavoured to lay and foster a rational foundation. The Indian cosmic view was primarily concerned with the cosmic *energetic* source as a positive unitary principle and as the 'womb' of all the created manifestations. There was, in addition, a conception of the Light-Infinite (*aditi*), as the infinite substratum of all that is here and beyond. Even the Vedic gods were regarded as the manifestations of Infinite Light. In the *Upaniṣads*, a series of philosophical compositions, which expound the ultimate reality and the relation between the self and the Universal self, the Brahman is referred to as the 'Self-Luminous Light, the Light of lights and the Divine light before which thousands of Suns pale into nothingness'. Thus, the ancient Indian perspective encompassed a transcendental unitary energetic principle as the Primordial Being, of which the celestial luminaries – the Sun, the Moon, the planets and the stars – were the manifest components.

Two facets

The Indian man-spirit-cosmos view has two notable facets. First the eternal or imperishable human spirit, in contradistinction to the transient or perishable human body, constantly endeavouring to be in harmony with the cosmos, pursuing the avowed goal of merging with it eventually; the second is the cyclic concept. The former represents an innate attitude of the self towards all that is non-self, a spiritual communion of the microcosm with the macrocosm or Nature, as opposed to the exploitation of Nature for material benefits. The latter concept is of far more significance inasmuch as the cyclic happenings in the celestial sphere are reflected in the human socio-religious life. The recurring changes in the position of the Sun during the day, the periodic waxing and waning of the Moon, the regular appearance and disappearance of the planets, the night and day as well as the seasons, the birth and death all unmistakably projected a cycle of events which gradually were woven into the socio-religious fabric. Further the celestial cycles and the recurrent celestial phenomena were used to identify auspicious moments for the performance of various rites.

In this respect, the Vedic notion of sacrifice (*yajña*) is very important, since it was conceived as the navel of the universe and as the unifying principle between Man and the Universe. The Vedic seers believed that *yajña* is the pathway for preserving and keeping in constant movement the cycle of events – both microcosmic and macrocosmic – in a harmonious matrix of space and time. They even conceived it as a connective dynamic principle in the manifested world as a protector and nourisher of the Universe. The most important consideration was that the sacrifice, to be effective, had to be performed at auspicious times and within the altars of exact geometrical forms propitiating therein, the fire – a formless agency of communion of man with the cosmos. The form and the formless, together with the human spirit, were supposed to become divine.

Thus, the astronomical endeavours in India drew their first inspiration from the general Indian ethos of an integrated man-spirit-cosmos view. The principal activity was the time-reckoning and the calendrical computations in terms of the motions of the Sun, the Moon and the planets along the zodiacal path with which were also associated 27 or 28 asterisms. It is for this reason, perhaps, that one does not notice in Indian astronomy any special inclination to reflect upon the origin, nature and structure of the celestial bodies, nor any effort in mapping of stars outside of the zodiacal path. In other words, astronomy in India had a rather limited objective; yet within this framework, it began to take certain strides of considerable magnitude.

Vedic astronomy

On the basis of well established evidence the origins of Indian astronomy can be definitely traced to the Vedic period (*c.* 1500 B.C.). In the Vedic literature *Jyotiṣa*, which connotes 'astronomy' and which later began to encompass astrology, was one of the most important subjects of study, being the foremost among the auxiliaries of the Veda. The earliest Vedic astronomical text has the title, *Vedāṅga Jyotiṣa*, which avers that 'one learned in the

Vedas who has also learnt the lore of the movement of the Moon, the Sun and the stars, will enjoy, after death, a life in the world wherein the Moon, the Sun and the stars move and he will also have on the Earth an unending progeny'. A precise knowledge of the movements of the heavenly bodies, specially of the Sun and the Moon, of the eclipses and the solstices, was essential for the socio-religious life of the people, because this helped the determination of the time, both auspicious and inauspicious, for the performance of sacrifices, religious observances, festivities, marriages, and agricultural operations, on which perhaps depended the spiritual goal of the people, viz. an innate desire to be in communion with the cosmos, its beauty and harmony.

Apart from the details of various sacrificial rites and the specification of their performance at propitious times, we also find in the Vedic literature a specific view of the Universe in terms of three distinct aspects – the Earth, the Firmament and the Celestial. Of the celestial luminaries, the Sun was regarded as the most important and its ecliptic path, assiduously determined over a long period, was considered sacrosanct. There were several rituals associated with specific positions of the Sun along the ecliptic. Obviously, the Moon was the next most important luminary, specially for time-reckoning. In fact, the Sanskrit word *māsakrt*, for the Moon, means that the latter is the 'maker' of a month. It is interesting to note that the English word, month, is derived from 'mooneth', indicating the association of the Moon with a month. There were two systems of month-reckoning – one ending with the New Moon, and the other with the Full Moon.

The path of the Moon was observed in relation to 27 or 28 asterisms (called the *nakṣatras*), and the lunar zodiac was well determined. It may be noted that the Chinese (*hsiu*) and the Arabs (*manazil*) also had a similar system. Whether each civilisation developed this system independently of the other, or whether there was mutual borrowing or there was any common source for all, is still a moot point. Many of the Chinese *hsius* differ from the Indian *nakṣatras*; only nine of the 28 *hsiu* determinatives are found to be identical with the corresponding *junction stars* (*yogatārās*) of the concerned *nakṣatras*. On the other hand the Arabian *manazils* seem to be more or less in agreement with the Indian system. In any case, the method and the manner pursued by the Vedic priests for purposefully delineating the asterisms unmistakably reveal their originality. The names of the lunar months were given on the basis of the *nakṣatra* in which or near which the full Moon occurred. The twelve lunar months were generally divided into six seasons of two months each, and there were also special names for solar months (see Table 2.1).

The different Vedic calendars were: a sidereal lunar year (12 months of 27 days, or 13 months of 27 days each); a synodic lunar year (6 months of 30

Table 2.1. *Nakṣatras and their modern equivalents*

No.	Nakṣatra	(Yogatārā) Junction star
1	Aśvinī	Beta Arietis
2	Bharaṇī	Fortyone Arietis
3	Kṛttikā	Eta Tauri
4	Rohiṇī	Alpha Tauri
5	Mṛgaśiras	Lambda Orionis
6	Ārdrā	Alpha Orionis
7	Punarvasu	Beta Geminorum
8	Puṣya	Sigma Cancri
9	Aśleṣā	Alpha Cancri
10	Maghā	Alpha Leonis
11	Pūrva-Phalgunī	Delta Leonis
12	Uttara-Phalgunī	Beta Leonis
13	Hasta	Delta Corvi
14	Citrā	Alpha Virginis
15	Svātī	Alpha Bootis
16	Viśākhā	Alpha Librae
17	Anūrādhā	Delta Scorpii
18	Jyeṣṭhā	Alpha Scorpii
19	Mūla	Lambda Scorpii
20	Pūrvāṣāḍhā	Delta Sagittarii
21	Uttarāṣāḍhā	Alpha Sagittarii
22	Śravaṇā	Alpha Aquilae
23	Dhaniṣṭhā	Beta Delphini
24	Śatabhiṣj	Lambda Aquarii
25	Pūrva Bhādrapada	Alpha Pegasi
26	Uttara Bhādrapada	Gamma Pegasi
27	Revatī	Zeta Piscium

days and 6 months of 29 days each); a civil (Sāyana) year (12 months of 30 days each); and a pseudo-solsticial year of 378 days, by adding 18 days to the third year after two civil years of 360 days each, to bring about correspondence between the civil and the sidereal solar year of 366 days (i.e. the interval between two successive passages of the Sun in its apparent annual motion through the same point relative to the fixed stars). Intercalation was adopted for luni-solar adjustments.

There is enough evidence in the Vedic literature to confirm that the Vedic Indians possessed the knowledge of both the winter and summer solstices. In fact, a year was generally divided into two halves of six months each, marked by the winter and summer solstice respectively. A month was divided into two parts – the bright and the dark half of one lunation, each half consisting of 15 units (called the *tithis*) and this type of division is characteristically Indian, ingeniously devised for calendrical purposes. A day was regarded as consisting of 30 time-units (called the *muhūrtas*). The largest at the summer solstice comprised 18, and the shortest at the winter solstice, 12 such time-units[†]. Divisions of a year, a month and a day, practical as these were, served the socio-religious life of the people encompassing the annual, monthly or daily performance of sacrifices and other rites, festivals, religious vows and the like.

From an annual cycle to a conception of a cycle of five years was a step forward and we notice such an endeavour in the Vedic astronomical text, the *Vedāṅga-jyotiṣa* (probably 1400–1200 B.C.; in its present form, may be 700 or 600 B.C.). The quinquennial luni-solar cycle has been worked out here in considerable detail. The text also deals with intercalation of two months in the five-year cycle – one at the end of the fifth solstice and the other at the end of the tenth, for luni-solar adjustments. It also provides a rule for determining the length of the day between the two solstices. These astronomical parameters were effectively woven into the socio-religious life of the people.

The Siddhānta astronomy

The period between the time of the *Vedāṅga-jyotiṣa* and the 5th century A.D. when the first mathematics based astronomical text the *Āryabhaṭīya* appeared, is the one which, admittedly, received certain external influences, specially from the Greek and Greco-Roman regions. Greek astronomy itself had imbibed Babylonian influences. In any case, after the invasion of Alexander the Great and the subservient rule that followed, a certain flow of astronomical knowledge must have occurred and, in the process, the elements of Hellenistic astronomy were assimilated into Indian astronomy. Alongside, there was a considerable spurt in the local astronomical endeavours. The two or three centuries preceding and following the Christian era proved to be formative in India. It was during this period that new astronomical parameters specially of the planets were being determined, and a system of coordinates adopted. The *nakṣatra* system was gradually being replaced by the 12 signs of the zodiac. The length of the year was determined as accurately as possible, the planetary motions with geometric models of eccentric circles and epicycles were studied in detail, and the occurrence of solar as well as lunar eclipses was calculated. This period also provided a womb, as it were, for a new class of astronomical works, called the *siddhāntas* (literally 'final conclusion or solution') with appropriate technical terminology in Sanskrit. Moreover, a mathematical approach lay at the

---

[†] Editor's note: an interesting interpretation of this specific difference between the shortest and the longest days in a year has been given in the next article.

bottom of all these endeavours. From then on, competent Indian astronomers were also able mathematicians.

While a critical examination of the chronology of the major astronomical
works generally attributed to this period is a matter of detail the five *siddhāntas* (later summarized by Varāhmihira in the 6th century A.D.) merit special
attention. They are: the *Saura*, the *Paitāmaha*; the *Vāśiṣṭha*, the *Pauliśa* and
the *Romaka*. Of them, as Varāhamihira emphasizes, the *Saura* or the *Sūrya
Siddhānta* a text by an unknown author, is the most accurate. It underwent
improvement through modifications between the 6th and the 12th centuries
A.D. The extant *Sūryasiddhānta*, in its 14 chapters, deals with the mean
motion of the planets, their true positions, lunar and solar eclipses, planetary
conjunctions, junction-stars, heliacal risings and settings of planets, cosmog-
ony and geography, reckoning of time etc. as well as several astronomical
instruments – all in about 500 verses. An important aspect of this text, is the
conception of a huge cycle of 4 320 000 years (the *mahāyuga*) to which we
shall return later.

As to the other four *Siddhāntas*, the *Paitāmaha*, rather an inaccurate one,
gives astronomical constants similar to the *Vedāṅga-jyotiṣa*. The *Vāśiṣṭha*
specially deals with the true motions of the five planets, giving their synodic
periods as well as their equivalence to the sidereal revolutions. The sidereal
year length postulated in this text is 365 days. The *Pauliśa* reveals a knowledge of the anomaly and the equations of the centre, and one can observe
in this work the beginnings of spherical trigonometry.

There was an effective transmission of the knowledge from one generation
to another. The standardised terminology, contained in elegant verses, and
the novel methods of number-reckoning which were in linguistic tune with
the meters of the verses not only engendered the rationale of astronomy and
its associated mathematics but also provided new stimuli for the growth of
astronomical literature in the centuries that followed.

The astronomical literature mainly in Sanskrit can be broadly classified
into specific texts dealing with astronomical operations and solutions (the
Siddhāntas), computation rules and methods (the *Karaṇas*), and the astronomical tables (the *Koṣṭhakas*), besides those dealing specially with the astronomical instruments.

According to David Pingree, who has carried out an extensive survey of
Indian astronomical literature, there are in India and outside some 100 000
manuscripts on the various aspects of *jyotiḥśāstra*. Admittedly, the great
majority of them were copied and recopied during the last five centuries or
so, since manuscripts cannot survive long in Indian weather conditions. Even
presuming that 80 per cent of them might deal with astrology and other
pseudo-scientific aspects, there could be at least about 20 000 manuscripts on
Indian astronomy (the number of titles may be less) of which barely a couple

of hundreds have been critically studied so far. The wealth of manuscripts unmistakably reveals the prolonged astronomical tradition in India. This long tradition has produced a distinguished chain of mathematician-astronomers almost to the present day. It is wellnigh impossible to list all of them. Nevertheless, special mention needs to be made of Āryabhaṭa I (5th century A.D.); Varāhamihira (6th century); Bhāskara I and Brahmagupta (6th–7th century); Lalla and Govindasvāmin (8th–9th centuries); Vaṭeśvara, Muñjāla, Śrīpati and Āryabhaṭa II (10th century); Someśvara and Śatānanda (11th century); Bhāskarācārya II and Sūryadevayajvan (12th century A.D.); Āmarāja (13th century); Makkibhaṭṭa and Mādhava (14th century); Parameśvara, Yallaya and Nīlakaṇṭha (15th century); Śaṅkara and Acyuta Piśāraṭi (16th century); Viśvanātha and Putumana Somayāji (17th century); Maharaja Sawai Jai Singh II and Jagannātha (18th century); and Śaṅkaravarman (19th century). Their treatises are noted for a methodical treatment of several astronomical aspects as well as for the elegant literary forms engendered by an expertise in Sanskrit and an innovation in communication.

Main characteristics

At this stage, it is desirable to deal with certain characteristics of Indian astronomy, in a perspective. First, the geocentric and geostatic concept which dominated the thought of the Indian astronomers. The Earth was considered to be a stationary sphere at the centre of the solar system. The Sun, the Moon and the planets were regarded as having their own motion west to east, while the asterisms or the stellar spheres were considered to have their motion from east to west, as a result of which the former were supposed to fall behind the latter. However, still adhering to the geocentric concept, Āryabhaṭa I modified the geo-stationary view by postulating a rotation of the Earth about its axis and gave a precise rate of the rotation – a determination which is very close to the modern value. He explained the apparently retrograde or westward motion of the stationary asterisms. Unfortunately, he was the lone crusader of the concept of rotation of the Earth, and was strongly opposed by the succeeding generations of astronomers. In any case, the motions of the planets, the reasons for their very slow, slow, fast and very fast motions, the observed retrograde and transverse motions were all worked out within the geocentric model.

Next the Cyclic Concept. A reference has been made earlier to the huge cycle of 4 320 000 years called a *mahāyuga* (literally Great Epoch) which constituted the most fundamental element of Indian astronomy, specially from the 4th century A.D. onwards. There was even a period 1000 times longer, i.e., 4 320 000 000 years (*Sanskrit:Kalpa*). A notable aspect of the general Indian thinking in respect of what may be called the eternity of time, was that even the conceivable minutest interval of time, like the one taken for winking the eye lids, was related step by step, to the day, month, year, epochs and what was conceived as a Divine Year. In this manner, the microcosm was sought to be integrated with the macrocosm and the great cosmic cycle. Nevertheless, of particular interest from the standpoint of Indian astronomy is the period of 4 320 000 years at the beginning and end of which the planetary bodies were supposed to be in conjunction having undergone integral number of revolutions in between. The beginnings and the endings were only for practical determination of their positions, while they were regarded as moving endlessly in a cycle. There is no denying the fact that such a huge period for computing the mean motion of planets is peculiar to Indian astronomy. Perhaps Indian astronomers conceived of this huge period with a view to avoiding fractions concerning the number of intercalary months, omitted lunar days, civil days and the like.

There were some variations in the conception of the huge cycle. Varāhamihira's version of the *Sūrya Siddhānta* (in his *Pañcasiddhāntikā*) deals with a period of 180 000 years and, with reference to it, gives the number of the intercalary months as well as the omitted lunar days in whole numbers. If this number is multiplied by 24, we get 4 320 000 and the concerned astronomical elements could thus be worked out. Āryabhaṭa I divided the *mahāyuga* into four equal parts of 1 080 000 years each. The *Sūrya Siddhānta* has divided it into four ages: *Kṛta* or Golden Age; *Tretā* or Silver Age; *Dvāpara* or the Brazen Age; and *Kali*, the Iron Age – the English equivalents are only for purposes of understanding. The duration of these ages are in the descending order of 4:3:2:1. The number of solar years in the *Kṛta*, *Tretā*, *Dvāpara* and *Kali* would be 1 728 000; 1 296 000; 864 000; and 432 000 respectively. Through such nomenclatures, the astronomical texts endeavoured to be in tune with the traditional and even the mythological concepts. A notable feature of the Indian astronomical concept concerning the *mahāyuga*, or the commencement of special epochs, is that all the planets are supposed to be in conjunction at the initial point of the celestial sphere; in other words, their longitudes would be zero. Thus, if the number of civil days elapsed from the beginning of an epoch up to a desired time is determined, the mean longitude of any planet could be worked out. The Indian astronomers coined a term, *ahargaṇa*, to denote the number of civil days elapsed during any interval of time. Further, for luni-solar reckonings, the number of intercalary months were taken into consideration. The Indian astronomers developed also methods for determining the time, longitudes of the Sun and the Moon as well as those of the planets. As and when the necessity arose, they had even devised elegant and reasonably accurate methods for the corrections like the deficit of the Moon's equation of the Centre as also for the equation of time due to the obliquity of the ecliptic.

The celestial sphere was conceived in all its details – the zenith, the nadir, the horizon, the prime vertical, the hour circle, the meredian, the ecliptic, the celestial equator and the inclination of the ecliptic to it, the celestial poles, etc. In general, Indian astronomers, through calculations and models of their own, appeared to have arrived at a value of 24° for the obliquity of the ecliptic which formed the basis of their other astronomical computations. Yet another fundamental postulate was that the first point of the asterism (*nakṣatra*) *Aśvinī* (near the star *Zeta Piscium* or *Revatī*) was regarded as the fixed point from which the longitudes were measured.

Precession of equinoxes

In the ancient times, there were divergent views on the observed phenomenon of the retrograde motion of equinoctial points on the ecliptic, which altered the coordinates of the stars in various ways. Now, we know that this is due to precession of the Earth's axis with a period of about 25 800 years.

In India, there was, for quite some time, the concept of the libration of equinoxes, i.e., an oscillation about the fixed point of *Aśvinī*, the maximum eastward or westward deviation being 27, as envisaged in the *Sūrya Siddhānta*. The text states that there are 600 such to and fro oscillations in a cycle of 4 320 000 years, the period of one such oscillation thus being 7200 years (annual rate of precession would work out to be 54″). Leading astronomers like Āryabhaṭa I, Brahmagupta and Lalla did not seem to have thought of the precession of equinoxes. The commentator of the *Āryabhaṭīya*, Bhāskara I, was even averse to such a concept. Varāhamihira, though well aware of the phenomenon, did not show any evidence of his knowledge of the rate of precession.

In the late 7th century, one Devācārya was perhaps the first astronomer who conceived of a method for computing precession. Later astronomers like Vaṭeśvara and Āryabhaṭa II evolved methods of their own, even so in terms of solsticial points. In the 10th Century A.D., the astronomer Muñjāla recognized the precessional motion, in terms of complete cycles and stated that the equinoxes would revolve 199 699 times in 1000 *mahāyugas*, which works out to an annual precessional rate of 59″.9. Another astronomer, Pṛthūdakasvāmin, worked out the rate of precession as 56″.82 per year.

Eclipses

For nearly two thousand years, the mythological view of *Rāhu* (presumed to be the head of a demon) devouring the Sun or the Moon thereby causing eclipses, held sway over the minds of priestly astronomers who were also astrologers. However, it was Āryabhaṭa I who, in the 5th century A.D., provided an explanation for the eclipses in terms of the Sun being obscured by the Moon, and the shadow of the Earth obscuring the Moon. Varāhamihira explained clearly the cause of a lunar eclipse as being due to the entry of the Moon into the shadow of the Earth. Nevertheless, he preferred to use the nomenclature of *Rāhu*, stating that the ascending node is *Rāhu's* head while the descending node is *Rāhu's* tail. Later the word *pāta* began to be used for connoting the nodes. It would seem that some astronomical texts also used the words *Rāhu* and *Ketu* in the technical sense of ascending and descending nodes. As to the parallax, the Indian astronomers tried to evolve methods for the parallax correction, specially for calculation of eclipses.

Certain rites and rituals during eclipses have been an integral part of Indian religious life, then as now, and the occurrence of eclipses was awaited with great religious fervour. The Indian astronomers gave due importance not only to the detailed calculations of the occurrence and the duration of the solar and lunar eclipses but also to observing them diligently. For instance, astronomer Parameśvara (15th century A.D.) observed and even recorded the lunar and solar eclipses which occurred over a period of fifty years. There are also inscriptional records of eclipses which are generally concerned with certain propitiatory acts like gifts made by rulers and chieftains on such occasions. Such records, however, have not yet been examined from the astronomical point of view.

Instruments

The oft-repeated question, 'did Indian astronomers employ instruments for observing the motion of the heavenly bodies and recording them?', needs some consideration. It may be noted that a substantial number of the major astronomical texts contain chapters on instruments, their making and methods of using. Āryabhaṭa I has emphasized the importance of careful observation. The various instruments fabricated and used by the astronomers included shadow instruments, rotating spheres including the armilliary sphere, water instruments, circle and semi-circle instruments, scissor- and needle-types, graduated tube, rectangular plate-like contrivances and the like. Some of the texts on astronomy have given clear instructions as to how they can be used effectively specially for determining the time, cardinal directions, altitude and zenith distance, hour angles, conjunction of planets and the like. Unfortunately, however, so far no attempts have been made to prepare models of these instruments and examine them to find out about their accuracy.

The medieval period

The astrolabe began to be used in India during the medieval period, as a versatile instrument. Most of the astrolabes, now preserved in the Museums in India, belong to the 17th–19th centuries and the engravings on them are in Persian characters, though a few of them are engraved in *Nāgarī* characters. There was even a family of astrolabe-makers in Lahore. There is no denying the fact that the astrolabe represents the impact of Arabic astronomy on India in the medieval period.

Fig. 2.1. An Indian astrolabe of the late medieval period. Courtesy: Red Fort Museum, Archaeological Survey of India, New Delhi.

India, by virtue of its geographical position, of its links with trade routes, both ancient and medieval, and of its cultural contacts, played its own part in the transmission and assimilation of scientific ideas, more so since the advent of the commercial or trade relations with Ptolemaic Egypt, Roman Empire and West Asian Caliphate. During the reign of Caliph Al-Mansur, an Indian astronomer visited Baghdad and transmitted some of the Indian methods of computations, as evidenced by the account of Ibn al-Adamī in his astronomical tables entitled *Nazm al-igd*. The Indian texts – the *Brāhma-sphuṭa Siddhānta* and the *Khaṇḍakhādyaka* of Brahmagupta – were rendered into Arabic by Muhammed ibn Ibrahim al-Fazāri and Ya'qūb ibn Tariq under the titles *Sindhind* and *Zij-al-Arkand* respectively, in the 8th century A.D. Āryabhaṭa I was known to the Arab astronomers as 'Arajabahra'. The Central Asian and other Islamic astronomers like Al-Khwārizmī, Al-Hasan bin Misbah, Al-Nairizi, Ibn as-Saffar, Ibn Yunis and Al-Battani were quite familiar with Indian astronomical computations and even used them in their works. Not a few of the Islamic *Zijes* or astronomical tables, included the zero meridian of Ujjain, under the name 'Arin', the *Kali* era, etc. From the Islamic culture-area came to India the most outstanding scientific transmitter and synthesiser, Al-Bīrūnī, in the 11th century A.D. His meticulous work, *Ta'rīkh al-Hind* was indeed a classic and a valuable source for Indian astronomy of the times. Such was the catholic attitude of the savants of the ancient and medieval times. They travelled far and wide in search of knowledge possessed by people belonging to other regions of the world.

In the medieval period there emerged a considerable volume of astronomical literature in Sanskrit, but the majority of them was in the nature of secondary works and commentaries on the earlier works like the *Sūrya Siddhānta*, and of Āryabhaṭa, Brahmagupta, Lalla, Bhāskara I and Bhāskara II. However, some of the commentaries also included improved methods of astronomical calculations and tables. There were noted families of astronomers specially in Banaras, Maharashtra and Kerala, who assiduously fostered the astronomical tradition by developing succinct methods of calendrical computations and producing elaborate commentaries on the classical texts. Another notable dimension was added to the activity during the 16th–17th centuries, specially when the traditional Indian astronomers interacted with those who were in the Mughal Court. Several Arabic and Persian texts on astronomy and mathematics were translated into Sanskrit and studied by Indian scholars.

In this respect, special mention needs to be made of Maharaja Sawai Jai Singh II (1688–1743) of Jaipur, who was not only an able king but also a skilled astronomer and patron of learning. He built five observatories in different locations in Northern India. The observatories now standing majestic and serene in Jaipur and Delhi bear testimony to his abiding interest

Fig. 2.2. Misra Yantra viewed from the South, 'Jantar Mantar', Maharaja Jai
Singh's Observatory in Delhi. Courtesy of Archaeological Survey of
India, New Delhi.

in astronomy and to his efforts for augmenting the astronomical tradition
with an open-mindedness. The observatory at Jaipur has a large number of
instruments – huge sun-dials, hemispherical dial, meridian circle, a graduated
meridianal arc, sextants, zodiacal complex, a circular protractor (which are
masonry instruments) as well as huge astrolabes. Sawai Jai Singh II meticu-
lously studied the Hindu, Arabic and the European systems of astronomy.
He was well aware of Ptolemy's *Almagest* (in its Arabic version), as also the
works of Central Asian astronomers – Nasir al-Din aṭ-Ṭūsi, Al-Gurgāni,
Jamshid Kāshi and, more importantly, of Ulugh Bek – the builder of the
Samarqand observatory. In fact, it was the Samarqand school of astronomy
that appears to have been a great source of inspiration to Jai Singh in his
astronomical endeavours.

No less was his interest in European astronomy. In his court was a French
Jesuit missionary who was an able astronomer and whom Jai Singh sent to
Europe to procure for him some of the important contemporary European
works on astronomy. He studied Flamsteed's *Historia Coelestis Britannica*,
La Hire's *Tabulae Astronomicae* and other works. He was well aware of the
use of telescope in Europe and he spared no efforts in having small telescopes

Fig. 2.3. Rasi Valaya, 'Jantar Mantar', Maharaja Jai Singh's Observatory in Jaipur. Courtesy of Archaeological Survey of India, New Delhi.

constructed in his own city. In the introduction to his *magnum opus, Zīj Muhammad Shāhī*, which is preserved both in Persian and in Sanskrit, he has recorded that telescopes were being constructed during his lifetime and that he did make use of a telescope for observing the sun-spots, the four moons of Jupiter, phases of Mercury and Venus, etc. However, in the absence of a critical evaluation of his treatise, it is rather difficult to opine whether Jai Singh was able to determine the planetary positions or movements with the help of a telescope and whether he recorded them. No positive evidence has yet been unearthed.

The principal court astronomer of Jai Singh II was Jagannatha who was not only well versed in Arabic and Persian but also a profound scholar of Hindu astronomy. He translated Ptolemy's *Almagest* and Euclid's *Elements* from their Arabic versions into Sanskrit. The *Samrāṭ Siddhānta*, the Sanskrit title of the *Almagest*, is indeed a glorious example of the open mindedness and generous scientific attitude of Indian astronomers.

New trends

In the 16th–18th century medieval India, there was yet another influence on the astronomical endeavours through the Jesuits who came from Europe and who had the expertise in geography and cartography in terms of practical astronomy. Father Anthony Monserrate (1536–1600), Father Jean-Venant Bouchet (1655–1732), Father Richaud (1633–1693), Father Boudier (1687–1707) who worked in Jai Singh's observatory at Jaipur, Father Figueredo (1700–1753) who, at the request of Jai Singh, went to Europe and brought back with him some of the European works on astronomy, and the other missionaries provided new dimensions to astronomy in India.

The East India Company too was encouraging astronomical endeavours and helped in the establishment of a number of modern type of observatories. Of them, the Madras Observatory, which came into being in 1792 at Madras, the forerunner of the later Kodaikanal Observatory, deserves special mention*. John Goldingham, Warren, T.G. Taylor, N.R. Pogson and others, not only equipped this observatory with the then available modern instruments but also conducted several investigations. The first Indian astronomer who made a mark in this observatory was one C. Ragoonathachari who was associated with some 38 000 observations in the *Madras Catalogue of Stars*, besides discovering two new variable stars (*R Reticuli* and *V Cephei*). He was elected a Fellow of the Royal Astronomical Society in 1872.

The Indian endeavours such as those indicated in the preceding paragraphs amply demonstrate that, since there was a long astronomical tradition in India fortified by an open mindedness, western astronomy was not looked upon as something alien to the Indian ethos, but, instead, it soon found Indian adherents and a congenial home. As a result of this attitude, India could nurture such intellectuals as M.N. Saha and Vainu Bappu who made outstanding contributions to modern astronomy, enlarging our perception of the Universe. It should be said to the credit of the Indian culture-area that it continues to nurture its ancient astronomy as a living tradition and at the same time endeavours to foster the modern astronomical investigation. For, who knows exactly as to which of the two, provides a better understanding of the intricate relationship between the microcosm and the macrocosm?

---

* Editor's note: There has been recent research to show that Madras Observatory was set up in 1786 as a private observatory by William Petrie, an Englishman. In 1790, the observatory was taken over by the East India Company with one Michael Topping as the Company Astronomer. The year 1792 mentioned in the text is actually the year when the observatory building came up at Madras Nūngambakkam. (See Kochhar, R.K. 1985, *Bull. Astr. Soc. India* **13**, 162).

3

Making of astronomy in ancient India

DEBIPRASAD CHATTOPADHYAYA

A masterly analysis of global archaeological data led Gordon Childe to the concept of the 'Urban Revolution'. Urban revolution refers to the profound socio-economic transformation that ushered in the earliest cities in human history. Childe recognised three primary centers of this in the ancient world, viz. Egypt, Mesopotamia (the present day Iraq) and India (primarily the Indus Basin). One of the profoundest achievements of this revolution, in Childe's view, was the birth and growth, in these cultures, of 'exact and predictive sciences'. Both the Egyptian and Mesopotamian civilisations had established traditions in mathematics and astronomy and had evolved suitable calendrical systems. These are well documented since papyri and clay tablets of these civilisations have been successfully deciphered. In contrast, the Indus Valley Civilisation (also called Harappan Civilisation) has been notorious for the difficulties it has presented to the historian in the deciphering of its records. In spite of numerous claims to the contrary, the Harappan 'seals' remain undeciphered and hence inaccessible to date. Thus we lack direct documentary evidence of the development of astronomy in this civilisation. Yet the main purpose of this essay is to argue in favour of the view that astronomy started in India in the pre-Vedic era with the Indus Civilisation.

The earliest Indian literary compositions are called the Vedas – the *Ṛgveda*, *Sāmaveda*, *Atharvaveda* and *Yajurveda* – which in total bulk are literally staggering. Though it is impossible to be exact about their dates, it is generally believed that these were orally composed by a branch of the Indo–European language speaking people, who called themselves Aryans and who entered India roughly in the 16th century B.C. as predominantly pastoral nomads. Of the four Vedas we shall be referring particularly to the *Yajurveda* which was later than the *Ṛgveda* and which took a sharp turn to discuss the rituals. It appears that there were once over a hundred recensions of the text, of which only a few have reached us. These are roughly dated as 1000 B.C. Apart from the detailed prescriptions for the performance of the rituals, the *Yajurveda* contains theological or quasi-theological discussions that foreshadow the next phase of development of the Vedic literature, which are called the *Brāhmaṇas*. According to the modern scholars, the *Brāhmaṇas* date from somewhere between the 10th and 7th centuries B.C., when the Aryans moved from the north–west where they had entered India and started settling in the Indo-Gangetic divide and the Upper Gangetic Basin – geographically somewhat below the latitude 28°N. Of the *Brāhmaṇas* we shall be referring mainly to one, which is known as the *Śatapatha Brāhmaṇa*.

Specially in these ritual texts we come across glimpses of some calendrical system and, therefore, also of some astronomical views on which it is based since these texts proposed to specify the time, day, season or some observed celestial phenomenon appropriate for the performance of the rituals. Among the modern scholars, B.G. Tilak and H. Jacobi, independently, proposed to revise the generally accepted date of the Vedic literature on the basis of these astronomical data. They recovered from the Vedic literature certain references to basic astronomical observations. Analysis of these by modern methods indicated a hoary antiquity – the 3rd or 4th millennium B.C., for their actual occurrence. According to Tilak the period could be even earlier. These two scholars, then, argued that the Vedic literature itself was of the same antiquity since they assumed that the observations referred to in the literature were from the same period as the literature itself. Such an assumption is generally questionable and particularly so in the present case for a number of reasons.

First, the astronomical data found in these Vedic texts are so desultory and so deeply embedded in discussions concerning ritual trivialities and theological disputations that it is nearly impossible to locate them and then disentangle the data from the framework of highly quaint logic which sought support in these. It is indeed difficult to imagine that the authors of these texts had any genuine interest in astronomy which they seem to have used for sheer mystery-mongering. The priests, whom we meet in the *Yajurveda* and the *Brāhmaṇas*, interested as they were above all in their *dakṣiṇā* or the sacrificial fee, could, quite conceivably, be trying to use every scrap of astronomical data that they had inherited from antiquity, to add to their own rituals an awe-inspiring appearance without ever bothering to verify these by direct observations. This belief is strengthened when one sees that the astronomical observations referred to in the *Brāhmaṇas* do not quite fit in with the spatial and temporal contexts of these texts. Secondly, there are other sounder ways of dating the Vedic literature (e.g. philological considerations) and these disagree with Tilak and Jacobi on pushing back the dates of the Vedas to the hoary antiquity. It is outside the scope of the present discussion to consider the technicalities of the question of dating Vedas but we can still quote some authoritative opinions on this. R.S. Sharma, the noted Indian historian states that: 'The French scholar Louis Renou, a lifelong student of the Vedic texts, accepted the view of Max Müller that the Aryans appeared in India around the 15th and 16th centuries B.C. and placed the hymns of the *Ṛgveda* around this date.' The *Yajurveda* texts are of a much later origin. Sharma continues: 'On the present showing the use of iron in the Indo–Gangetic divide and the Upper Gangetic basin, in which the *Yajus* texts and the *Brāhmaṇas* and *Upaniṣads* were compiled, cannot be taken back earlier than 1000 B.C., for this metal is known to several texts. Renou thinks that the *Brāhmaṇas* should be placed between the 10th and 7th centuries B.C.'

Keeping these views in mind, we may now examine some of the data of interest found in the Vedic literature. One of these concerns a certain asterism – *Nakṣatra* (usually rendered as 'lunar mansion' or 'lunar station') – called *Kṛttikā*. The Vedas as well as the different recensions of the *Yajurveda*, e.g. the *Taittirīya-saṃhitā*, *Kāthaka-saṃhitā* and *Maitrāyaṇī-saṃhitā* give a list of twenty seven *nakṣatras* beginning invariably with *Kṛttikā*. Jacobi argued that *Kṛttika* was assigned the first position because during the period of these texts, its position coincided with the vernal equinox. This led him to conclude that the Vedic culture was already in existence by 3000–2000 B.C. While his last inference is open to question, Jacobi was certainly right about the position of *Kṛttikā*, as some further remarks made about the asterism in these texts show.

One such remark is to be found in a section of the *Śatapatha Brāhmaṇa* where it forms part of a theological controversy. The controversy is concerning the setting up of the two fires–called Gārhapatya and Āhavanīya under the *nakṣatra* considered most auspicious for the purpose. The opposing views are regarding whether the fires were to be set up or not under the *Kṛttikās*. The dispute itself is of little interest to the historian of science. However, the following fact mentioned in the course of the argument is of scientific relevance – it states that the *Kṛttikās* do not move away from the eastern quarter while the other asterisms do so. To quote the text

> *etā ha vai prācyai diśo na cyavante sarvāni ha vā anyāni nakṣatrāni prācyai diśas cyavante.*

Commenting on the words *prācyai diśo na cyavante* (does not swerve from the East), Sāyaṇa, by far the most authentic commentator of the Vedic texts, observed *niyamena śuddhaprācyām eva udyanti* i.e. 'rises invariably in the due east'. Scholars have accepted this distinctive property of the *Kṛttikās* as observed then.

Based on astronomical treatise like the (modern) *Sūryasiddhānta*, *Kṛttikās* are identified with Pleiades in Taurus, with Eta Tauri (Alcyone) as its determinative star. If, therefore, we accept this as an actual piece of observation codified in the *Śatapatha Brāhmaṇa*, it is possible to determine, with the help of the methods of spherical astronomy, the date when this observation was made since the present position of Eta Tauri in the sky as well as the precessional rate of the Earth's axis are also known to us. It turns out that the date when the position of Eta Tauri coincided with vernal equinox is as early as 2334 B.C.[†] On the same basis in 1000 B.C., the earliest time that could be assigned to the *Śatapatha Brāhmaṇa* by responsible Vedic scholars, the celestial longitude of Eta Tauri would be 18°30′. Thus the statement that the *Kṛttikas* rose exactly in the east could by no means be based on an actual observation from the period of the *Śatapatha Brāhmaṇa*. On the other hand, it is entirely possible that the piece of observation that the *Kṛttikas* rose exactly in the east formed part of the astronomical knowledge of a much earlier period and came down to the authors of the *Śatapatha Brāhmaṇa* who accepted and codified it without verification. The priests whom we encounter in the *Brāhmaṇas* show such utter lack of genuine scientific interest that it is not surprising they never bothered to find out if during their time the *Kṛttikas* rose in the due east or not.

Once we accept this historical interpretation we are led to speculations with far-reaching consequences. A people, genuinely interested in astronomical observations, would not be satisfied with noting just an isolated phenomenon like the rising of *Kṛttika* in the east; it is more logical to presume that they would have a whole system of astronomy, of which this observation formed a part. If such a system were from a culture antedating the period of the Vedas, it is possible that the Vedic priests used scraps of knowledge from it whenever the need arose, mainly to settle their nonscientific disputes. Could it be then that the entire *nakṣatra* system which we find in the *Saṃhitās* and the *Brāhmaṇas* had its real roots in the astronomy of a culture which preceded the Vedic?

In fact, it is the analysis of the astronomical data found in the Vedas which led Tilak and Jacobi to the conclusion that the Vedas actually far antedated their then accepted date of birth. What they ignored was the possibility that the actual observations were not contemporaneous with their codification, that they possibly formed part of a tradition which came down from a hoary antiquity only to be used in codified form at a much later period by a people who had little enthusiasm for direct observation or lacked an adequate scientific temper to verify what they inherited. That the Vedic priests generally encouraged speculation rather than direct observation is amply evident in their writings. As they put it

paroksa-priyah iva hi devah pratyaksa-dvisah

[†] *Editor's note*: That Pleiades are among the first stars mentioned, appearing in the Chinese annals of 2357 B.C., Alcyone then being near the vernal equinox (see *Star names – their lore and meaning* by R.H. Allen, Dover, 1963), is in remarkable agreement with the view expressed in this article.

i.e. 'the gods are fond of deliberate mystification, and they detest direct observation'.

It would be unfair on Tilak and Jacobi to not mention here that during their time nothing was known about any Indian civilisation predating the Vedas; for them the Vedas were the necessary starting point of understanding Indian culture. Jacobi understood this limitation in his own way when he said in 1909 that if 'we are quite sure that Vedic culture was not older than 1200 or 1500 B.C.' we would be obliged to seek other explanations of the astronomical data contained in these. Referring to the usually accepted date of the Vedas he added: 'As long as this fact remains in suspense, either my arguments or these three subversive interpretations given to them by my opponents will appear plausible in accordance with the estimated age which critics assign to Vedic culture. When the new theory on the antiquity of the Veda was first discussed, I made this same statement to Mr Tilak, who wished to enter upon a campaign against all opponents. I told him that the discussion would have no definite result unless excavations in ancient sites in India should bring forth unmistakable evidence of the enormous antiquity of Indian civilisation.'

The discovery of the Indus Valley Civilisation within a decade and a half of Jacobi's prescient remarks settled this question once for all. The story of this discovery, though well known, does deserve a few lines here. It was in 1921 and 1922 that preliminary diggings at Harappa by Daya Ram Sahni and at Mohenjodaro by R.D. Banerjee yielded identical finds including exotic seals and soon the potential of the sites came to be realised and the elements of a forgotten civilisation identified. After examining the collection of antiquities from these two widely separated sites and being convinced that they were totally distinct from anything previously known in India, John Marshall announced the discovery in 1924. This took the archaeological world by surprise, for the excavations at Ur by Leonard Woolley, almost during the same period, had already created a great sensation. Marshall remarked that 'the discoveries had at a single bound taken our knowledge of Indian civilisation some 3000 years earlier'.

The relics of the Indus Valley Civilisation are found over a vast area covering roughly 80 000 square miles, with at least fifteen cities, so far unearthed, each containing a population of about 5000. The estimated population of the bigger ones like Mohenjodaro varies between 35 000 to over 40 000. There is no doubt that the civilisation thrived mainly on agricultural surplus and the recent discovery of a ploughed field at Kalibangan, with marks of ploughed furrows, leaves us with little doubt about the advanced agricultural technique then in use. Such advanced agriculture specially in a region always confronted with the problems of flood and inundation makes a strong case for the knowledge of some calendrical system, and, hence, of astronomy in certain form. Another remarkable feature of the civilisation is the meticulous town-planning uniformly followed in all the cities unearthed and this is evident in the 'rigid north–south and east–west orientation of the streets and lanes'. In addition, in the burials the dead were placed with their heads pointing to the north and the feet pointing to the south. Both these features indicate that the people had knowledge of stars in some form. It is further claimed that this civilisation was in communication with Mesopotamia through maritime trade and this would imply, once more, knowledge of the sky for purposes of navigation specially in default of the discovery of any compass or its prototype.

Highly sketchy and stray, such are some of the material evidences of the existence of astronomical knowledge in the Indus Valley Civilisation. To these may be added the findings from another line of research which is increasingly gaining prominence. Two teams of scientists, one in the Soviet Union and the other in Scandinavia, have been involved in efforts to decipher the Indus script using computers. Their results converge on the claim that the inscriptions on the 'seals' are pointers to a calendrical system. The Indian scholar I. Mahadevan has been trying laboriously to work out the same thing in his own way. Though it is somewhat premature to report on the final outcome of these efforts, it is proper to mention at least one point here. The calendrical system claimed to have been revealed in the inscriptions is same as or similar to what is called the 'sixty-year cycle (the cycle of Jupiter)' – a cycle of five twelve-year periods with some zoomorphic symbol to indicate each year. Among the neighbouring countries, Tibet had borrowed this calendrical system. The Tibetan scholars persistently claim that the system originated in some place called Sambhala, which remains to be identified but which could be somewhere within the cultural zone of the Indus Civilisation.

An alternative way of dealing with historical problems of the kind posed by the Indus Civilisation, where the lack or inaccessibility of the direct records makes any analysis sketchy and conjectural, is to apply the 'method of retrospective probing' on which I propose to put special emphasis. The assumption on which this is based may first be mentioned. When we come across some data codified in the literature of a period, which on various considerations, seems to be much later than the period of the data itself, the possibility of moving backward to some period, when such data are logically conceivable, should be seriously considered. The presumption then is that although the data were contained in the literature of a later period, their origin is to be traced to an earlier period where they were generated. Jacobi and Tilak followed such a procedure unconsciously and somewhat erroneously when they analysed the astronomical data found in the Vedas and opened for us the possibility of determining the antiquity of Indian astronomy. In the present article we have already given an example of this method when we discussed the possibility that the *Kṛttikas* rising in the due east far antedated the codification of it in the *Brāhmaṇas*. To continue further with this, we note that if the later Vedic literature contained data, which appear to be a chronological pointer to the Harappan period, they also contain data which may be taken as a geographical pointer to the Harappan region as the place where the astronomy in question began. In the following we shall concentrate on one such datum. This is to be found in *Vedāṅga Jyotiṣa*, meaning literally, 'astronomy as a limb of the Vedas', which claims to codify the views of a certain Lagadha, about whom nothing is known other than the fact that the name is clearly nonVedic. The datum in question is part of a description of the annual motion of the Sun. According to *Vedāṅga Jyotiṣa*, one solar year consists of 366 civil days, a civil day being the time from one sunrise to the next, i.e. nychthmeron of the Greeks. The solar year was divided into two equal *ayanas*, each consisting of 183 civil days. They were *uttarāyaṇa* and *dakṣiṇāyana*. *Uttarāyaṇa* defined the period of the northward movement of the sun from its southern-most declination i.e. beginning with the winter solstice to the summer solstice and *dakṣiṇāyana* defined the period of its reverse movement (southward) i.e. beginning with the summer solstice to the winter solstice. The text asserts that during the *uttarāyaṇa* the day increases by one *prastha* and the night decreases by the same amount while during the *dakṣiṇāyana* just the opposite takes place. Finally it states that the maximum time difference between a day and a night is 6 *muhūrtas* which is also the time difference between the shortest and the longest day in a year. The *Vedāṅga Jyotiṣa* goes on to describe the various units of time mentioned above. A certain contraption called *jala-yantra* was used for time measurement. Nothing but a vessel with a hole, the water discharged from it was measured in units called *prastha*. Without going into the details of the measurement and the definition of the various units of time, it should suffice for us to know that one *muhūrta* equals our 48 minutes and one civil day is thirty *muhūrtas* long, to be exact. Thus, according to *Vedāṅga Jyotiṣa*, on summer solstice when the day is the longest and the night the shortest, the time difference between them is six *muhūrtas* or 288 minutes. That is, the day is 18 *muhūrtas* long (our 14 hours 24 minutes) and the night lasts 12 *muhūrtas* (our 9 hours 36 minutes). Having described the crucial datum, we can now state our inference. Since the relation between the shortest and the longest day of the year is a function of the geographical latitude of the place of observation, the particular time difference between the two mentioned in the *Vedāṅga Jyotiṣa* should immediately tell us the latitude of the place where the observation was made. A simple exercise in spherical astronomy leads to the answer that the latitude is 34°5′ N. Since the Vedic people eventually settled and produced their literature much further south (approximately 28° N), the astronomical contents of the *Vedāṅga Jyotiṣa* could not be based on their own observations. The northern latitude inferred for the place of observation falls within the cultural frontier of the Indus Valley Civilisation. Thus we have here another example of the astronomical data of the Harappans which came down to the Vedic priests and were codified in their literature and perhaps used by them without verification.

The observation under discussion of the longest and shortest day specifies only the latitude of the place of observation but not the longitude. As a result it could also be that the datum was from the Mesopotamian Civilisation, whose southern capital Babylonia was located at 32°5′ N. This perhaps is one of the many reasons why it has often been suggested that Indians borrowed their astronomy from the Mesopotamians. I differ from this point of view. Apart from the possibility of independent and parallel developments in two different cultures, we should also note that the *Vedāṅga Jyotiṣa* view is based on detailed calculation and also on the use of a certain apparatus for measuring the time-unit. We do not find parallel calculations, nor the relevant instrument for measuring time-unit in the Mesopotamian culture. Mere coincidence in the geographical latitude and the recognition of the fact that Mesopotamians also had developed their system of astronomy do not justify such a sweeping conclusion. Both Whitney and Thibaut dealt with this point. Although their remarks were made long before the discovery of the Indus Valley Civilisation and therefore quite obsolete in the light of our present knowledge, it is worth considering them since by these remarks they may have inadvertently perpetuated certain misconceptions about the achievements of the Aryans. Thus Thibaut wrote in 1877: 'Regarding the disputed point whether the rule fixing the length of the shortest and longest day of the year has been borrowed by the Indians from some foreign source, for instance from Babylon, or sprung up independently on Indian soil, I am entirely of the opinion of Prof. Whitney who sees no sufficient reason for supposing the rule to be an imported one. It is true that the rule agrees with the facts only for the extreme north-west corner of India; but it is approximately true for a much greater part of India, and that an ancient rule – which the rule in question doubtless is – agrees best with the actual circumstances existing in the north west of India is after all just what we should expect.' Since for Thibaut the only valid starting point of Indian history was the *Rgveda* and since it was unanimously assumed that the Vedic people entered India from the north-west, he seems to have tacitly assumed that while entering India from the north-west, these people actually observed that the longest day consisted of *18 muhūrtas* and the shortest of *12 muhūrtas*. However, there are a number of reasons why we cannot agree with Thibaut's assumptions. If the Vedic people actually observed such a phenomenon while they were entering India from the north-west, it is only reasonable for us to expect some reference to it in the earliest stratum of the *Rgveda*. But there is nothing in the entire *Rgveda* even remotely suggesting this. Secondly, when the Aryans entered India, they were on the whole nomadic pastoral people whose socio-economic life did not need astronomical knowledge. When they settled down many centuries later in the '*Aryavarta*' (or '*Madhyadesa*') and converted to agriculturists, astronomical knowledge perhaps did become indispensable for them though from what we have said it appears that they depended for this purpose more on surviving heresay than on actual observations. Thirdly, from what we read about the technological development of the Vedic people in the *Rgveda* itself, it is difficult to imagine that during the period of the oral composition of this vast literature they could improvise the time-measuring instrument which was so central to the determination of the lengths of the day and night.

With the benefit of hindsight we could now say that the Harappan Civilisation, which flourished in north-west India at least a thousand years before the arrival of the Aryans, did require and depend upon astronomical knowledge for its socio-economic development producing the vast agricultural surplus for which evidence exists. It is left to the future historians to decipher the records of this civilisation and confirm what we have asserted here about the existence of a system of astronomy in this civilisation. In the absence of direct documentary evidence we had to follow the 'method of retrospective probing' and have tried to show that the literature of the Vedic people does contain data which in all probability originated in the astronomy of this earlier culture.

4

The impact of astronomy on the development of western science

JEAN-CLAUDE PECKER

The famous line from J.-M. de Heredia: '... Du fond de l'Océan, des étoiles nouvelles'[†] reminds us of the incredible effect astronomical discoveries have always had upon Earth's exploration. That it is an aid to navigation, that it is necessary for measuring the Earth and was until recently required for an exact determination of time, all this is well-known. But an even more profound result of astronomical discoveries has been the recognition of man's smallness in the universe and of the unity, if not also the universality, of science. We are today entering a new era in which astronomy can not only lead us to a vaster and even more magnificent laboratory for exploring our solar system, but will also open entirely new perspectives for the future of mankind. In this essay, I examine the impact of astronomical discoveries on the development of western science. It is an honor to have been asked to do my part in this tribute to the memory of a great scientist who was also a humanist. But I do so with some sadness, too, for Vainu Bappu was my friend, as well as my colleague. We had similar views about the role of astronomy in the modern world, and agreed in particular on the need – an ethical need – for the scientist to make a concrete effort to *explain* the universe to the rest of society. I will try here, then, to present what I can of the ideas and beliefs we shared.

My understanding of one important area is far from adequate to fulfil this duty, and Vainu would have been the one to teach me a great deal of what I lack for presenting the importance and influence of early Indian astronomy. Indeed, in all those parts of the world not directly influenced by the Greco-Roman culture of the so-called West, astronomy was highly developed and was a source of considerable practical application. Besides Indian, one must also recall the importance of astronomy in the Chinese, Mayan, Incan and African empires.

[†] '... From the depths of the ocean, new stars.'

However important these cultures and however great the place of astronomy within them, one must nonetheless recognize that in the history of science and, even, of society at large, their scientific developments were progressively forgotten and replaced by western science, be it for the good or for the bad. Chinese astronomers, as only one example, knew about supernovae long before Tycho; but it is Tycho's discovery (in 1572) that changed astrophysical concepts, not the Chinese ones made centuries earlier. So, however regrettable it may be, we cannot ignore that the modern world has been influenced almost entirely by science as it developed in the western world.

For centuries the sky offered primitive men and women a reference point in the turmoil and confusion of their hazardous lives. After all, the seasons rule the course of life, just as day and night succeed each other with unwaning regularity. From the sky came not only the succoring gifts of the Sun but also unavoidable catastrophes; they quite naturally placed their gods, both good and bad, into the sky, and toward it directed both their hopes and their fears. The sky and its features became a true source of communication between peoples, so that it is hardly surprising to discover how many of the ancient myths were linked with the seasons, with the Sun and the Moon, with eclipses and constellations, and how many similarities both hidden and obvious exist between the myths of various cultures. No wonder, then, that the knowledge of heavenly bodies – of their appearance and behavior, their disappearances and returns – was felt to be a natural need and a means to a better acquaintance with the divine universe as much as with the physical one.

But was this already astronomy? If one looks at the results, one must agree that it was. Eclipses were predicted at a surprisingly early date, planetary motions known, and calendars established. And prediction of the seasons was an obvious need for early agricultural development.

A clear distinction was made between, on the one hand, the nearby meteors, whose behavior seemed almost human in their changeableness and sensitivity to extreme and unpredictable climatic changes in the heavens and, on the other, the impassible and permanent upper heavens of the myths and gods.

Once man started to travel and explore the Earth in a methodical way, the observations and meditations collected earlier were both immediately useful and the basis for gaining more knowledge. The Greek explorers, for example, were already able to use timekeeping devices and astronomy to map the Mediterranean area and orient themselves with the help of the sky while navigating. And until the last century, sailors were still depending on the methods Christopher Columbus and Magellan used, to discover unknown parts of the world and to circumnavigate the sphere. With sextants and clocks and with astronomical knowledge, they were able to determine the height of stars above the horizon and to tell the direction of the geographical pole whenever the sky was clear. Nor were they fooled by magnets that indicate only the local magnetic properties of a field which is only roughly comparable to that of a dipole, while the dipole itself differs notably from the Earth's polar axis. Magnets were used only during bad weather and were known to

Fig. 4.1. *To the shepherds of all times, the starry night is a guide, as well as a clock.* (After Camille Flammarion, Astronomie populaire.)

be unreliable long before Jules Verne's tales of villains who led ships astray by the misleading use of magnets (in: *Un capitaine de quinze ans*).

Astronomy has from the earliest history been used to measure not only the Earth but also time. All methods of determining time have been linked first with the Earth's daily rotation and with its revolution around the Sun. Thus were the earliest calendars established and time scales defined: sundials in Roman times, the pyramids before them, the slow yearly round dance of the

Fig. 4.2. *Kronos, the God of time*: a clear astronomical symbol, which illustrates the importance of astronomy for the definition of time. (After Camille Flammarion, Astronomie populaire.)

Sun through the zodiacal belt that marks the seasonal succession. Only in the last decade were the astronomical definitions replaced by an atomic one, and then only after the present accuracy of measurement made us aware of the irregularities in the Earth's motion owing to the seasonal growth of plants or to tidal frictions. Perhaps even this latest perfect clock, the atom, will some day be found wanting when we detect irregularities based on some as yet unpredictable phenomenon – quarks or some unknown – that might provide a method of measuring time more accurately than atomic oscillations.

Beyond these last two applications one can mention two sciences closely associated with astronomy, so that their practical applications are in some way also tied to the study of the sky. Thus geodesy and astrometry as they developed during the seventeenth and eighteenth centuries included the measurement of Earth's properties, such as its shape and motion. The Académie des Sciences de Paris organized several expeditions to measure the shape of the Earth in 1735 and 1736, one to South America with La Condamine and Bouguer and one to Lapland with Maupertuis and Clairaut. They concluded the Earth was flattened at the poles, which then offered a convincing argument in favor of Newtonian gravitational theory. Later, the passage of Venus in front of the Sun was observed from several observatories in 1761 and 1769, and the data led Lalande to a precise estimate of the distance between the Sun and the Earth. All this was, of course, a continuation of Eratosthenes' efforts around 200 B.C. when he measured the radius of the Earth for the first time; and progress in accuracy has not stopped since the eighteenth century.

From the quotidian and terrestrial applications of astronomy mentioned above, one moves easily into describing the great advances in measuring the universe itself. Instruments have become more and more powerful, allowing greater penetration of even greater distances within the universe. And the further we explore those enormous distances, the further we move into the past, for a quasar ten billion light years away gives us information about the universe at that same location ten billion years ago, not at the present epoch of our own observing. And what we learn from the evidence filtered down through the ages has grown as well from star counts and descriptions of the objects' morphology to diagnostic analyses of the light itself with the use of photometers, spectrographs, polarimeters, etc.

But for all the uncertainty one would expect from information gained with such faint pencils of light, we have been able to evaluate temperature, pressure, density, and chemical composition in all points of the observable universe and from these to determine the object's stage of evolution. We now know there is a universal hierarchy: from dust and planets to stars, from stars themselves to star clusters and thence to galaxies, from galaxies to groups or clusters of galaxies, and lastly to the superclusters, and the observable universe. We therefore have gained considerable knowledge of space, where objects of all kinds are evolving, where one can observe their birth in their birthplace, or where one can see their death at the end of a turbulent development: the history of the universe, in short.

This growth in the understanding of the universe has had the specifically philosophical consequence of changing man's view of himself as part of his surroundings. In the immensity of the universe – and remember that our knowledge of its farthest reaches is relatively recent – man is a minute and modest being who may be able to express numerically this immensity but hardly able to conceive it, so large is the contrast between his own local and restricted preoccupations and the actual realm of the galaxies.

Each gain in knowledge has meant a move toward a new, less anthropomorphic view of man's place in the universe. How small the scale of the primitive universe was is clear from the early feelings of being equal to the gods, of being able to modify the course of events through prayer, of being able to fully perceive the universe with human eyes, and of being the center of justification of creation. Copernicus, and later Bradley, led to our putting aside this privileged position.

Meanwhile, not even the Earth is all that central to the solar system; and the Sun itself appears to be a modest and suburban star in our Galaxy. Even our Galaxy belongs to the 'local group' a group of galaxies near the outer border of the supercluster to which we belong (the Supergalaxy). And we now have reason to believe that our supercluster has no particularly privileged position in the universe. On a universal scale, then, we might well be considered accessories of no importance, or as Pascal expressed it so superbly, man is able to measure his loneliness between the two infinities of the microscopic and the gigantic universe:

> Que l'homme contemple donc la nature entière dans sa haute et pleine majesté; qu'il éloigne sa vue des objets bas qui l'environnent ... Que l'homme étant revenu à soi, considère ce qu'il est au prix de ce qui est; qu'il se regarde comme égaré dans ce canton détourné de la nature; et que dans ce petit cachot où il se trouve logé, j'entends l'univers, il apprenne à estimer la terre, les royaumes, les villes et soi-même son juste prix. Qu'est-ce qu'un homme dans l'infini?
>
> Mais pour lui présenter un autre prodige aussi étonnant, qu'il recherche dans ce qu'il connait les choses les plus délicates ...; qu'il se perde dans ces merveilles, aussi étonnantes dans leur petitesse que les autres par leur étendue ...
>
> Car enfin qu'est-ce que l'homme dans la nature? Un néant à l'égard de l'infini, un tout à l'égard du néant, un milieu entre rien en tout. Infiniment éloigné de comprendre les extrêmes, la fin des choses et leur

principe sont pour lui invinciblement cachés dans un secret impénét-
rable, également incapable de voir le néant d'où il est tiré, et l'infini où
il est englouti. . .

Le silence éternel de ces espaces infinis m'effraie.

B. Pascal, *Les Pensées*[†]

Neither the center nor the purpose of the universe, man remains an
accident in an evolutionary process that involves all matter from the oldest
days of the universe.

It is remarkable that, despite his limited possibilities for comprehending
the universe, man has nevertheless been able to construct a coherent physics
that can account for all observed phenomena. The universality of the laws of
physics is now considered essential to man's comprehension of the world, and
there is good reason to be proud of what has been acquired over such a short
span of time with but human skills. And there is reason to be more optimistic
even when discouraged by the other, more immediate challenges in this
difficult world and by the seeming inability of those same human skills to
manage and contain human aggression.

But then this last comment might well be used to claim that, apart from the
early sailors and the rare specialists in time determination, astronomical
research has been unable to change much in the human condition; so that the
philosophical justification we put forward before appears a poor expression
of personal preference; and our attempt at optimism based on gains in
astronomy seems a mere excuse for continuing to muse over quasars, to
theorize about flaring stars and to continue observing the infrared radiations
from distant molecular clouds.

[†] Let man contemplate the whole of nature in her full and grand majesty, and
turn his vision from the low objects which surround him ...

Returning to himself, let man consider what he is in comparison with all
existence; let him regard himself as lost in this remote corner of nature; and
from the little cell in which he finds himself lodged, I mean the universe, let
him estimate at their true value the earth, kingdoms, cities, and himself.
What is a man in the Infinite?

But to show him another prodigy equally astonishing, let him examine
the most delicate things he knows ... Let him lose himself in wonders as
amazing in their littleness as the others in their vastness ...

For in fact what is a man in nature? A nothing in comparison with the
Infinite, an All in comparison with Nothings, a mean between nothing and
everything. Since he is infinitely removed from comprehending the
extremes, the end of things and their beginning are hopelessly hidden from
him in an impenetrable secret; he is equally incapable of seeing the Nothing
from which he was made, and the Infinite in which he is swallowed up.

The eternal silence of these infinite spaces frightens me.

B. Pascal, *The Thoughts*

Such a claim would be false, for the universe remains a remarkable laboratory in which all is interrelated, and without astronomy to guide us there, our understanding of the physical world and the nature of modern life on Earth would be quite different and much more restricted. And how much the study of the stars affected man's understanding of life is well demonstrated by the major steps in the understanding of physical laws.

How all-encompassing the laws of physics are has not always been so obvious, for Aristotelian ideas assigned clearly different natures to the ether and the sublunar world. They saw the heavens as a pure element not subject to the humors of living beings and placed it in obvious contradistinction to the sublunar areas where the various combinations of fire, air, earth and water led to the complications and confusions inherent in life, death, putrefaction and catastrophe. In the eyes of the Greeks and of many of their medieval followers, the sky was then the seat of order, perfection and eternity. Earthly laws were different in essence from the celestial ones, and it would have been fearful indeed to consider they might be one and the same. The heavens therefore took part in the divine, and whole symbol systems

Fig. 4.3. *The mechanism of the heavens,* as seen by a medieval monk, discovering the physical laws behind the appearances. (After Camille Flammarion, Météorologie Populaire.)

were developed to express the final submission of the human and earthly to this divine, upper universe: the Great Chain of Being, the astrological macrocosm–microcosm correspondence and the various cosmological myths.

One can easily date a major turning point away from this philosophy on the bright Danish night of 11 November 1572, when Tycho discovered a star that was brighter than any other and that had not existed the day before. This star is known today as Tycho Brahe's supernova. The natural response at the time to such erratic stellar behaviour was to assign it a position in the sublunar medium. But this did not coincide with Tycho's perception of its distance and placement. Any object closer than the Moon would move more quickly in the sky with respect to the known stellar constellations and more quickly than the Moon. So in search of greater accuracy and understanding, Tycho tried to detect some movement of the star; but, for more than a year, the star, which grew weaker and weaker before disappearing to the naked eye, stayed in exactly the same place in the sky very near one of the five stars that constitute the W-shaped constellation of Cassiopeia. This meant that the new star had to be not only further away than the Moon, but also further than the Sun. And its brightness, after the first violent burst, changed regularly, thereby resembling all too closely for such a celestial object the earthly turpitudes it was supposed to rule above. Several years later, Tycho then determined the distance of a comet, and showed it to be several times more distant than the Moon. After this, who could rely on the Aristotelian dichotomy?

Tycho's discovery was only the first step in understanding that the laws of physics govern both realms and that we need only discover them, whether in the heavens or on earth, in order to understand our universe. Galileo and Fabricius showed, by proving that the spots visible on the Sun are indeed part of its surface, that even that most idealized of heavenly bodies was subject to phenomena that resembled anger, illness, temperament, even dirtiness. The distance of the Sun from the Earth was then estimated in the eighteenth century by studying over the surface of Earth the passage of Venus in front of the Sun; and this made it clear that the Sun was only one among billions of stars. Stellar distances were only measurable after about 1830 when Bessel, Henderson and Struve simultaneously (but independently) determined the distance of three stars. They used a trigonometric method. The difference in brightness between stars appears more as a function of stars' distance from the viewer than as resulting from any intrinsic difference between them. The Sun, for example, is similar in absolute brightness to the star Rigil Kentaurus, the brightest star in the constellation Centaurus, but the Sun is considerably brighter directly in relation to its closeness to us.

Even with these discoveries, the heavenly bodies were still considered by many philosophers mere points in the sky whose true physical nature could never be known. They were in a sense protected from human 'intrusion' by their distance, for the only information arriving from them were the light rays that had travelled vast, unimaginable infinities of time and space. This attitude was typical of the strict positivistic views of August Comte, among others:

> toute recherche astronomique doit être finalement réductible à une observation visuelle. Les astres sont ainsi, de tous les êtres naturels, ceux que nous pouvons connaître sous les rapports les moins variés. Nous concevons la possibilité de déterminer leurs formes, leur distance, leur grandeur et leurs mouvements; tandis que nous ne saurions jamais étudier par aucun moyen leur composition chimique,

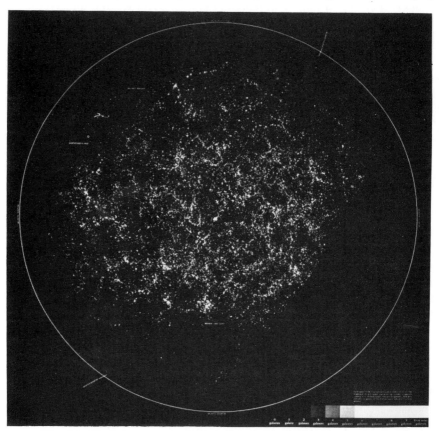

Fig. 4.4. *The universe: our ideal laboratory for physicists, chemistry and biologists.* (Here, after Peebles and Brand, a map of about one million galaxies of the northern hemisphere, brighter than magnitude 19. From one side of the map to the other, the light has to travel more than one billion years!)

ou leur structure minéralogique, et, à plus forte raison, la nature des corps organisés qui vivent à leur surface. Nos connaissances positives par rapport aux astres sont nécessairement limitées à leurs seuls phénomènes géométriques et mécaniques, sans pouvoir nullement embrasser les autres recherches physiques, chimiques, physiologiques, et même sociales, que comportent les êtres accessibles à tous nos divers moyens d'observation . . ..

Auguste Comte, Cours de philosophie positive†

Spectral analysis of these faint points of light changed the whole picture when developed in the mid-19th century, following Fraunhofer, Kirchhoff, and others. It began with the recognition that some elements present in earthly objects were also present in the Sun, e.g. hydrogen or iron. Progress was slowed when the spectral lines of what is now known as helium showed up in the solar spectrum – the Sun must be different after all! – but doubts subsided when helium was finally discovered both in rocks and in the Earth's atmosphere.

New doubts arose over the 'coronium lines' in the spectrum of the solar corona observed during total solar eclipses, and arose again because of 'nebulium lines' in the spectra of some nebulae, Orion's for one. The Mendeleev's Table continued to be filled without these two unknown elements being accounted for in any way. In 1919, Meghnad Saha discovered how to compute the level of ionization in any given atom; and this provided a breakthrough, for it became clear why no ion has the same spectrum as the atom from which it originated. Instead the resulting spectrum results from the interaction of the atom's nucleus with the electronic cloud surrounding it. Now with the aid of the Mendeleev's Table, completed by Boltzmann's law of excitation, Saha's law of ionization, and with the knowledge of the atomic structure and of the theory of atomic spectra, practically all spectra could be identified. Nebulium was nothing more than oxygen ionized twice, and coronium iron ionized nine and thirteen times. Now no element and no

---

† Any astronomical research should finally be reducible into a visual observation. Of all the natural beings, the stars are those that we can understand in the least varied respects. We conceive the possibility of determining their shapes, distance, size and then movements; whereas we shall never be able to study, by any method, their chemical composition or their mineralogical structure and all the more, the nature of the organised bodies which live on their surface. Our positive knowledge with respect to stars is necessarily limited to their geometric and mechanical phenomena, without being able to take up the physical, chemical, physiological and even social research which are involved in the beings accessible to our various means of observations.

Auguste Comte (*Lectures of Positive Philosophy*)

spectral line of any importance is still unidentified; almost all of the (more than) 100 000 lines in the solar spectrum are accounted for.

Added to the rich spectra of the stars, which are hot enough that atoms are ionized, are the spectra of colder bodies such as planets, comets and cold stars and of the interstellar medium itself. These lines not only appear in visible light, but also in the infrared and radio wavelengths. Radioastronomy at millimetric and centimetric wavelengths has, for instance, led to the discovery of hundreds of new molecules in interstellar and circumstellar matter.

Now not only spectral lines can be assigned to elements also known on Earth, but also their relative proportions in the heavens can be determined with a degree of accuracy no less than that of metallurgic analysis in any laboratory.

With all this, it has now been determined that the chemical composition of the universe is almost homogeneous as regards nuclei. Admittedly, differences in density and temperature induce spectral differences, as do the local effects of magnetic or electric fields, and the differential action of gravitation on heavy and light atoms; but all these can be explained with modern physics and chemistry. Cosmic organic chemistry is also similar to terrestrial organic chemistry; the interstellar medium contains formic acid, methyl and ethyl alcohols and amino acids, just like those on Earth. There is no reason to imagine that if life exists elsewhere, it would be chemically different than the carbon-based one we know. Speculation of phosphorus chemistry leading to some form of life is hardly supported by analysis of the various stars and nebulae.

Evidence is so strong to support the laws of physics that it would require the discovery of some totally unimaginable and unpredictable first principle of the universe to contradict them. It is therefore justified to apply the laws of the laboratory to heavenly bodies, and alternately to use the universe as a vast laboratory for exploring the Earth. After all, it offers such a variety of conditions, an immensity and a duration hardly matched on Earth and far from reproducible in any laboratory. It is also a relatively safe laboratory, for within it man is a passive observer. Only the latest space research has included human beings and thereby created 'experimental' astronomy. Where else was it safe to observe billions of thermonuclear explosions? And the analysis of flares on the Sun and the stars has given plasma physics information, without the need for experiments dangerous for others along with the scientists.

The relationship of astrophysics to physics itself illustrates our thesis of mutual dependence within the universe, for it has been the source of a great deal of knowledge in that field. A familiar branch of physics since Newton has been gravitational theory, expressed by a simple law that can be treated mathematically in a sufficiently easy and remarkably accurate way. Celestial mechanics is, in fact, based upon the application of Newton's gravitational law to celestial bodies; and among its many successes, one can count the accurate prediction of the 1759 passage of Halley's comet by Clairaut and Lalande. It was also used to accurately measure such basic constants as the astronomical unit of length based upon the distance between the Earth and the Sun, as well as to understand the apparent motions of satellites and planets and to describe stellar motions. With it, scientists could determine masses of objects in the solar system and of members of stellar binary systems.

One particular success for celestial mechanics based upon Newton's law was the discovery of planets as yet undetected, especially the discovery of Neptune by LeVerrier and Galle. Mercury, however, did not follow Newtonian computations, even when assuming the effects of local perturbations, as had been done when explaining the irregularities of Uranus's motions as the effect of Neptune's mass. Some astronomers posited an inframercurian planet to fulfil this role, but it was in vain. Only with the introduction of Einstein's General Relativity could one account completely for the advance of Mercury's perihelion.

Even though astronomy did not lead to the new gravitational theory, it did offer the most convincing arguments for its validity. Before Eddington analyzed the eclipse measurements of stellar deflection, as they were observed in 1919, General Relativity had been nothing more than the genteel hobby of an obscure physicist. From that day, Einstein became world famous – the Newton of our century – and, even more significant, the theory was then applied to other basic problems in physics, as well as in astronomy. It was used to explain Mercury's odd orbit, the bending of light rays near the solar limb, and the redshift of solar spectral lines in the Sun's gravitational field.

Thermonuclear reactions is another field which has benefitted considerably from astrophysics, while physicists have offered theories that have aided astrophysicists to understand this field as applied to the universe, most notably when trying to explain the Sun. Geological considerations made it clear that the Sun is very old, but it was unclear how the observed amount of solar energy could be continuously produced over such a long period of time. Chemical energy was suggested, but the more exothermic chemical reactions would have consumed the whole Sun in less than a few centuries. Then, in 1854, Helmholtz had the bright idea that contraction determined by the power of self-gravitation would convert the potential energy first into mechanical energy, then into thermal energy and lastly into radiant energy. But for all its ingeniousness, these computations only gave the solar mass a lifetime of a few dozen million years, while everything from geology and even the paleontology of the day spoke of a much greater age.

The problem could only be solved after another major discovery by Einstein, the equivalence of mass and energy as known through $E = Mc^2$, and after Becquerel and the Curies discovered the natural transmutation of elements. In 1919, J. Perrin was the first to find the correct answer when he explained the Sun's energy output by the transmutation of four hydrogen nuclei into a helium nucleus. The difference in mass is converted into energy according to Einstein's equation, which then allowed for a perfectly adequate solar lifetime that would continue for many billions of years.

This was only the first breakthrough, for much was left to explain about the solar machinery. Why, for instance, is this energy released at the rate it is, and why without any explosion, yet with enough power to explain the observed luminosity? In other words, the details of all the subsequent reactions had to be explained, taking into account collisional cross-sections, the lifetimes of unstable nuclei, and much more. Bethe and Von Weizsäcker finally succeeded in elaborating the reactions and, in doing so, introduced the so-called carbon cycle, in which carbon is the catalyst in the chain of thermonuclear reactions. It was proved later that the chain of proton-proton reactions was more appropriate to the solar situation than the carbon cycle.

We cannot go into all the details, but it is at any rate clear that solar studies have led physicists to laboratory work intended to elaborate the rate of nuclear reactions and to understand them better. And from this the idea of using nuclear energy came quite naturally, whether for the good or the bad. The bad results are obvious, no matter how one views the major powers' nuclear policy; but its use in non-military situations has helped solve some energy shortages in countries like France with little or no oil or gas reserves.

None of these uses are ever without hazard, for besides the dangers inherent in explosions and leaks, we have yet to find safe ways to solve the permanent problems of waste disposal, of dissemination on a world scale and of any military use of the non-military plants' energy. To solve the problem of any future energy shortage, however, we may need only continue the process of discovery into mastering fusion energy. We have thus far only mastered fission energy, but no fusion energy which is now only explosive and therefore only suitable for military uses. But the Sun's is fusion energy, so that once again we need to use our solar laboratory to help us find a technical solution to our present and future energy problems.

To get this far, it is necessary to understand plasma confinement and plasma physics is another field clearly aided by solar study as shown by the birth of what is now known as magnetohydrodynamics, or MHD. In the nineteenth century it was shown that the motion of matter is determined either by magnetic fields or by mechanical forces, but rarely by both at the same time, even when both are present. On a laboratory scale, the mechanical energy contained in one given volume is either much larger or much smaller than the magnetic energy so that either one or the other is the dominating force. But this is not true in astronomical plasmas, so that the Maxwell electromagnetic equations must be solved together with the equations of hydrodynamics.

This coupling is intricate, and the problems are compounded, as the following example will show. At the solar surface, forces (not very well known so far) expel matter into a zone where magnetic fields predominate. When matter leaves the solar surface, density decreases roughly as r^{-2}, and does so even more slowly in the streamers. Meanwhile, the magnetic field decreases faster; at less than one solar radius, magnetic energy has become so small with respect to mechanical energy that matter is no longer under its influence. The magnetic shield is destroyed, and matter goes its own way dominated only by the pure laws of hydrodynamics. Nevertheless, this transition zone is an important physical region where several solar features are created; and here a combined treatment of all equations is especially necessary.

To solve these kinds of problems, Alfvén created MHD in 1950. It is an important part of theoretical developments that may lead in the future to the mastering of plasmas and to the long-awaited control of fusion energy. The Sun's center offers a perfect example of what we have not yet been able to duplicate properly in any laboratory.

We could cite many other examples. Among them are the laser and maser, two discoveries that were strongly inspired by astrophysical studies of non-equilibrium thermodynamics and that have aided both modern physics and modern medicine: Think of retina detachment and other illnesses that can be treated by proper use of laser surgery! And we could continue with other examples not so familiar to the public. All together they emphasize how astrophysics is, by its methods, a part of physics; but then, physics itself, because of its need for varying conditions, is intimately related to astrophysics and greatly benefits from its development.

How better to study the universal laws of physics at work than with space observatories? With their use, the astronomical spectrum has been extended to include X-rays, gamma rays, ultraviolet and infrared wavelengths. And we have no doubt progressed in an understanding of active, even very violent, phenomena in the universe, such as the nuclei of active galaxies, which may teach physicists a lot about non-equilibrium situations. It may be too early to predict exactly what will come out of these developments; however, one cannot avoid thinking that the better we understand these very extreme situations in the universe, the closer we shall be to finding out what may have happened billions of years ago. At those times, densities and temperatures were such as to create conditions strongly dependent on microphysics; that is, the study of the 'big bang', if it existed, or at least an understanding of the highly condensed stages of matter, might offer clues to one of the greatest physical questions of all times: the 'grand unification' of all interactions from gravitational force to strong nuclear interaction.

The last decade has seen a new era in astronomical space research. Instead of being a passive science limited to observing phenomena without influencing it, space research has begun to allow astronomical studies *in situ*. The astronomical motivation for exploring the Moon or Mars may not be the main one – one should not underestimate the geopolitical source of space exploration, or its military sides – but it has also truly been a boost for pure science. Direct exploration of the planets has revealed unpredictable features: volcanoes on Io, whirlpools on Jupiter and Saturn, sulfuric acid in Venus's atmosphere, sand storms on Mars, and icebergs of ethane on Titan. One suspects these discoveries are only the beginning. What looks at first like astronomy has therefore benefitted the Earth sciences by widening the field of investigation to the surface and even the core of solid heavenly bodies. It can be predicted that new technological developments will come also from the mineralogical use of asteroid metallic content, as was suggested decades ago by Lyman Spitzer, one of the scientists of our day who goes furthest towards presenting the future of space science. What may now almost seem science-fiction, if rewritten in 2000, would contain many more proposals for applying space astronomy, while these might very well have been fulfilled.

Most of society can admittedly live well, watching television and eating quietly in families, with no knowledge of astronomy whatsoever. An overdose of television might even lead them to believe in astrology and to maintain only some vague, prescientific ideas about astronomy. It may well be true that we now know less astronomy than primitive man, who used the sky to fulfil agricultural duties, since meteorological services exist throughout the world and since calendars have long been perfected. But behind this domestic scene stand physicists and industrialists building this cosy envelope for the layman, and they are using more and more the discoveries that were obtained with the help of astronomical methods and investigations and the discoveries that concern ever remoter parts of the universe.

We arrive, therefore, at a paradox that might be dangerous if one let it continue in the direction it seems to be going. Should we aim at a society similar to the one envisioned by Aldous Huxley in *Brave New World*: superior men using their knowledge to build a comfortable cage for people with less knowledge? Abhorrent thought! To avoid such a development, and here I feel in full agreement with thoughts often expressed by Vainu Bappu, it seems necessary for scientists in general and for astrophysicists in particular to explain, explain, explain, without ceasing, and in simple but rigorous terms, what is going on in the heavens. This is the sacred duty of the scientist to society.

The author would like to thank Joli Adams for her indispensable help in editing the English text.

5

Man and the Universe

HUBERT REEVES

In this paper I will try to show how scientific discoveries have modified our vision of the universe throughout the centuries. 'Where am I? What am I doing here?', are the most natural questions that, since the dawn of history, people have asked, everywhere on our planet (and perhaps elsewhere). Science does not answer these questions. But it brings some elements of importance for personal investigation. These elements have evolved in recent centuries and in consequence our vision of the world has changed.

For clarity of discourse, I will distinguish three historical chapters on the question of 'how man situates himself in relation of the universe'. They could be called: (A) The ancient dialogue; (B) The universe does not answer any more; (C) A new vision is emerging.

(A) The ancient dialogue

In the quasi-totality of primitive cultures, as in most of the great traditional wisdoms, the Earth is the center of the universe. Above the Earth is the starry vault, the Heavens, where the Pantheons are situated. The universe is not very large. The stars and the planets are just out of the reach of our hands, but not out of the reach of our voice.

In the Heavens, live paternal figures who take care of human people in a very direct and personal way. Their homes are often on the top of the highest mountains, which are stairways from the Earth to Heavens. The human person does not suffer from loneliness, far from it. He is being watched all the time and must behave accordingly. Meteoric manifestations such as, thunder, lightning, comets, are all elements of a celestial language, often terrifying. The ancestors are up there. They have entered the Pantheon. They play the role of intermediaries between the outer world and the inhabitants of the Earth.

In many religious traditions, the divine figures are the parents of humanity. The relationship is more than simply symbolic, it is truly genetic. The event of the divine birth of human people is related in mythological texts. The relationship gives a meaning to human life. Man is held responsible for his destiny. By properly obeying the laws, he gets access to the world of the ancestors. Otherwise he is punished.

This presentation of our first historical chapter is, of course, highly simplistic. Such a vast subject could not be decently summed up in a few lines. Here I only want to point out a few salient features, common to many religious visions:

(1)   the world is small;
(2)   mankind is in constant contact with an outerworld;
(3)   humanity descends from divinity;
(4)   life has meaning and duties are clear.

(B) The universe does not answer any more

The astronomical shock

To our knowledge, the Ancient Greeks are the pioneers of scientific inquiry. They recognized the real nature and dimensions of the sky. To Heraclitus, claiming that the Sun is no larger than the foot of a man, Anaxagoras answered that it is probably larger than the Peloponnese ....

With the advent of the Renaissance, and the development of telescopes, the sky suddenly takes gigantic proportions. Our Earth is not the motionless pedestal around which all the stars revolve; *it is itself a heavenly body* just as the Moon and the planets. All these celestial objects move around the Sun which, at this moment, becomes the new center of the world.

Not for long. Soon one realizes that *our triumphant Sun is just another star*, like those which abound in the sky, on clear moonless nights. In fact our Sun is a very commonplace star, lost somewhere in the suburbs of our Milky Way. If the night stars are not as dazzling as the Sun, it is simply because they are so much farther (as already suggested by Anaxagoras). Light reaches us from the Sun in eight minutes; to travel from the nearest constellations, it takes many, many years ...

In 18th and 19th centuries, the Universe was identified with our Milky Way, or Galaxy, a volume extending over one hundred thousand light-years. Then, in the beginning of our century, a new revelation takes place. *Our galaxy is not unique.* New ones are constantly being discovered, more or less similar to ours. They are spread over billions and billions of light-years. Today, we count them by hundred of millions. We face seriously, in the framework of modern cosmology, the possibility that their number may be *infinite* and that the universe may be *literaly unbound*.

The feeling of dizziness faced with these new dimensions, and the realisation of our insignificance in the abyss of space, have seriously influenced the philosophical thinking of past centuries. The 17th-century French writer Blaise Pascal is for us, in this respect, a witness of prime value. As one of the most important scientists of his century, and at, the same time, a leading philosopher and theologian, he is ideally suited to appreciate the impact of these astronomical discoveries. His famous sentence 'The silence of these infinite spaces frightens me' most aptly summarizes the drama of this period. In this mute cold and empty universe, man is alone and alienated. Because of his deep religious faith, Pascal did not explore in depth this feeling of loneliness, as did later philosophers. Nevertheless he harshly acknowledged the blow dealt by the development of astronomy on the traditional views of the man–universe relationship.

The biological shock

A second shock was to come from the field of biological studies with the Darwinian theory of evolution. Here the genetic relationship between men and the outer world is being denied. Adam and Eve are no more created by God in the Garden of Eden, and the bright Athenians do not emerge, fully armored, from Jupiter's thigh. The ancestry is far less noble and glamourous. It is out of the vagina of apes that our ancestors came for the first time to the light of the day.

Moving back in time, in search of the ancestors of our ancestors, we meet progressively more primitive animals such as the reptiles, the fishes and minute amoebae swimming in the fetid water of faded flower vases.

To make good measure, the exploration of the fundamental mechanisms of biological evolution discloses the major role of *chance*. In the framework of natural selection, random processes are essentially involved in the emergence of new animal forms, which may or may not remain, amongst the realm of living species. This is often called the 'biological shock', a new blow to the Ancient Dialogue. Man is not the Son of God, rather the Son of chance. What can you say to chance?

The psychological shock

Two more blows were to hit the traditional man–God relationship during the 19th century, at a psychological level. The first from Marx, the second from Freud.

Exploring the deep layers of human psyche, disclosing the existence of the 'unconscious mind', Freud and his disciples alter considerably our perception of human will and responsibility. Handicapped by early conflicts, our margin of manoeuvre is far narrower than traditionally assumed. Furthermore, mental mechanisms are discovered by which the image of the father is projected beyond the family sphere, in the form of divine figures. Neither a genetic father, nor a spiritual father, God is seen as a product of infantile phantasmagoria.

Marxist thinking contests the image of God as the Supreme Chief of all nations. Social power and authority does not emanate from Heaven. The right to govern is, above all, the right of the powerful. The history of mankind is not the manifestation of a divine project but an anecdotical narration of the struggle of social classes.

One can easily understand the deeply depressive influence of these successive blows on the philosophical thinking of our times. They have generated what has been called the 'anguish of modern man' confronted with the silence of the Heavens, with its own solitude and with the absurdity of reality. Without contact with any 'beyond', alienated in his own world, western man occupies a position unique in the history of mankind; that of an essential psychotic.

(C) A new vision is emerging

We now came to the third chapter of our story. Here we present some features through which a new vision of the world is emerging.

In this vision, man is no more a stranger in the universe. Quite the contrary, he discovers that he belongs here, that he is inextricably involved in a long story which involves also the stars, the galaxies and universe as a whole. Ironically, this new vision emerge from the very same scientific facts which had given such lethal blows to the ancient dialogue, through a rereading and reappreciation of their meaning.

For many years, the various sciences have worked independently; each one in its own field of research, without any apparent contact or exchange with each other. Far from trying to widen their scope and to meet at the boundaries, these disciplines were getting more and more specialized, perpetually narrowing their interest, and concentrating on an ever smaller fraction of reality, to the point that a specialist of butterflies had no more exchanges with the specialist of spiders than with the volcanologist or the thermodynamicians .... Bernard Shaw has aptly grasped the situation in saying that if the man of culture knows 'nothing' on 'everything', the scientific expert knows 'everything' on 'nothing'.

It is in the present century that, thanks to the accumulated knowledge of the past, it becomes possible to achieve a synthesis of the various disciplines .... We discover now, what we had partly lost from view, that all sciences are studying the same object: the Universe, inhabited by man, the author of science. We discover that, by a judicious juxtaposition of all the scientific discoveries and results, we can draw a large fresco. Moving backward to get a general view, this fresco presents a central theme, a general web, in which the different branches of knowledge acquired by the various sciences, can be incorporated and related to each other, just as in a museum, when one walks back from a large painting to get a global view, one gets to perceive the leading patterns and movements which, from close-to, showed no coherence or rhythm.

The coordination of the various branches of scientific knowledge revolves around the general theme of the 'history of the universe'. Scientists of past centuries studied a reality which appeared eternal and fundamentally unchanging: the laws of the stars, the laws of the atoms. Gradually, the *historical dimension* of reality crept in and played an ever more important role. It began with geological studies of the past eras of our planet. Through the Darwinian theory of evolution, the 'historical' became an essential part of biology; animal species are continually changing and transforming from the early differentiation of primitive cells appearing four billion years ago. Astronomy, chemistry and physics join the 'wagon of history' with the discovery of the fact that atoms and stars have not always existed, that they come into being at certain moments, thanks to the combined interplay of gravitational, electromagnetic and nuclear forces.

More recently, cosmological discoveries have extended the historical vista to the whole of the universe. Not only the inhabitants of the universe change with passing time but also the very structure of the cosmos. The universe 'appears' fifteen billion years ago in a totally chaotic state of extreme heat and density. In these remote times, it harbours no structure, no organized system, no stars, no galaxies, even no molecules, no atoms, no atomic nuclei. It consists of isothermal and homogeneous 'soup' of 'elementary particles', that is, of simple elements which cannot be split into even simpler elements. Contemporary physicists would recognize the electrons, the photons, the quarks, the neutrinos, and a few other particles.

Modern sciences teach us how, chapter by chapter, matter has managed to organize itself as the temperature, gradually decreasing along with cosmic expansion, allowed the various forces of nature to come into play. *Astronomy* tells us of the formation of galaxies and stars by the effect of gravitational influences on the initial magma. *Physics* strives to elucidate the manner by which, under the spell of nuclear forces, elementary particles are fused into atomic nuclei in the fiery centers of stars. *Chemistry* reconstitutes the various phases in which nuclei and electrons, ejected from dying stars, combine, by the effect of the electromagnetic force, to form atoms, simple molecules and dust grains in the cold vastness of interstellar space. Back to work, gravity then gives birth to planets assembled from these grains. Some of these planets will have atmospheres and oceans, where chemistry, together with geology, tries to understand the formation of the first living cells. There, biology takes over joining forces with paleontology and archeology, and leads us along the trail of unnumerable biological species, all the way to the emergence of human beings, joining forces also with psychology and psychoanalysis in a common effort to understand human behaviour, including the development of scientific activity itself. Thus the scientist, far from being an observer of unchanging reality, finds himself becoming an *historian*, only to realize, at the end, that he is writing his own story, his *autobiography*.

The central theme of each one of these scientific chapters is the organization of matter. Science tells us that everything that exists: stones, stars, frogs, people, are made of the same matter, of the same elementary particles. The specific differences lie in the *state of organization* of this matter, of these particles with respect to each other. It lies in the number of steps climbed up in the scale of structuration. At the lowest level, the clouds of interstellar and intergalactic matter are in a state very close to the initial chaos. The stars are already somewhat higher in this respect. Gravity has shaped them in the form of spheres, but their level of organization is still very rudimentary. Stones are held together by the electromagnetic force. Embedded in a rigid crystalline network, the atoms of a stone manifest a level of organization far higher than the atoms of a star, but incomparably lower than the atoms of a frog. The human brain, with its billions of interconnected cells, constitutes the most formidably structured being known to us.

The performance of an organized system depends on its internal degree of complexity. Stars are content with shining, bacteria move around, feed and reproduce, but the highest performance ever accomplished in the history of the universe is the simple act of acknowledging our own existence together with the existence of the outside world. Science emerges from this activity of the thinking and questioning brain, in its effort to understand how he happens to be here.

Geographically, the quasi-totality of matter is endowed with a very rudimentary level of organization, akin to the early chaos. But in various privileged locations, on islands favored by appropriate conditions, matter has been allowed to yield to its natural tendencies toward organization, to bring forth the wonders of her ability.

We realize at this point how scientific discoveries and their integration in a coherent web lead us to reconsider the place of man in the universe, together with his relationship to reality. On the scale of mass and volume, man is nothing, a speck in an infinite space. But on the much more meaningful scale of organization, man occupies a high position. By our reckoning, he stands on the highest grade, that grade from which one can observe the universe and ask questions about its origin and future. No one before us (at least on our planet) had ever come to such interrogations.

From this observation-point, our relation to the universe is easily delineated. With the nebulae, the stars, the stones and the frogs, with everything that exists, we are involved and committed in that immense operation of matter organization. Contrary to the beliefs of past centuries, *we are no stranger to the universe*. We insert ourselves in an event which extends over billion of light years. We are the children of a universe which gave us birth after a gestation of fifteen billion years. In a truly Hinduisitic tradition 'the stones and stars are our brothers and sisters'.

How could the situation get inverted to that extent? How could science, after its repudiation of the ancient alliance between men and the cosmos, now present such a different vision of the world? To understand this transformation we shall consider again the various shocks described earlier. By appropriately integrating the recent scientific discoveries, we shall be lead to appreciate them in a different fashion. In short, we can say that these shocks have brought forward a more sensible view of the man–Universe relationship. They allow us to get rid of infantile images and reach a higher level of maturity, more on the level of the dimensions of our universe.

We consider first *the astronomical shock*, prompted by the discovery of the vastness of space. We realize today that this immensity is by no means a futile and useless luxury. In a smaller world, on the scale of archaic cosmogonies, matter could never have organized itself, even to reach the level of giant molecules. To generate atoms, we need first atomic nuclei which build up in the fiery cores of giant stars. To reach appropriate chemical abundances, we require generations and generations of stars, during billions of years, in galaxies similar to our Milky Way. Numberless stellar explosions have ejected and dispersed their harvests of fertile atoms in the vastness of interstellar space. Furthermore, we realize today that the very expansion of the universe is a condition essential to the advent of life on our Earth. First, because expansion is the motor of the cosmic cooling which allows the combination of atoms; in the initial heat, no stable system could permanently hold together. But also because, by constantly enlarging the gap between galaxies, the expansion takes care of the additional entropy produced in consequence of the material structuration, preventing it from threatening, by its accumulation, the bound system themselves. If the birth of life requires a gestation of many billion years, the expanding universe must consistently spread over billion of light-years. In short, the first shock has replaced a minute world full of anthropomorphic figures by a gigantic space, humming with the fever of cosmic gestation to which we owe our own existence.

Thanks to the *second shock*, the discovery of biological evolution, preceded by nuclear and chemical evolution, we have come to unravel the successive chapters of this cosmic gestation. Far from 'descending' from the gods, we are the result of a long 'ascent'. Nature has its own way of generating complexity, by association and different century. We emerge from a line of ancestors, in which we recognize, in turn, and in inverse chronological order, the apes, the mammals, the reptiles, the fishes, the cells, and still earlier the molecules, the atoms, the nuclei, the nucleons and finally the elementary particles of the big bang.

At the same time, we understand how organization could emerge from initial chaos. Between the particles of the cosmic soup, there are *forces* which will shape the matter when the diminishing temperature will allow it. Here, indeed, we shall meet '*chance*' but we shall interpret its presence in a new way. In the 19th century, the occurrence of random processes in nature was given as a proof of the absence of any kind of natural planning or project. In the cosmic-evolution scheme, chance is seen as playing *an essential role* in the actualization of the immense potentialities of matter. The richness and diversity of forms in the realm of the living could only come about in a world where the laws of physics do allow many different beings to exist and where these beings are brought into existence through the game of chance. The encounters of nuclei in stellar cores, the molecular captures in the early oceans, the impact of cosmic rays on chains of DNA in biological cells permanently create the new and the unimagined. Through these innumerable processes, nature finds it possible to manifest its wonderful potentialities.

Here we have to take account of a still largely mysterious aspect of the development of the world; the onset of *phases of disequilibrium*, essential to the advent of matter organization. The growth of natural complexity is under the constant threat of 'thermal death' which occurs when matter enters a regime of equilibrium (same temperature everywhere or same rate of particle capture and dissociation). If the universe was in such a regime, its appearance would be vastly different; it would be made entirely of iron nuclei (on the microscopic scale) and black holes (on the macroscopic scale). Cosmological studies show that the mode of universal expansion has successively brought matter into regimes of disequilibrium with respect to the reactions initiated by the various forces of nature: gravitational, chemical, nuclear. Not only is matter not in a state of equilibrium, but, as times goes on, it moves farther and farther away from such a regime. With respect to thermal death, we are in reprieve, a reprieve which progressively becomes longer and hence leaves ample time for more innovations and creations. There lies the source of freedom and inventiveness in nature. Why is the mode of expansion the way it is? We have a few ideas but we are far from being able to give satisfactory answers here.

In summary, we are the manifestation of the potentialities of an extremely inventive matter, whose actualization requires the intervention of chance and puts human beings in the progeny, not of divinities, but of apes. Matter behaves in a special way, whose finality is still unknown to us, but we are sure that if it behaved differently, we would not be here to realize it.

The psychological shocks have largely contributed to make our vision of the world more integrated, more mature, throwing light on those unconscious processes which are responsible for projecting on divine figures the actors of our inner conflicts. Notwithstanding these realizations, the Universe still remains, for us, deeply mysterious and full of marvels. In my opinion, the organizational potentialities of early matter may find an appropriate expression in the oriental concept of Tao, without figure, without personality, but ontologically preexisting. For a western expression, I would suggest the following verse of Charles Baudelaire:

> Nature is a temple where living pillars,
> Sometimes let go of confused utterances
> Man walks amidst forests of symbols,
> Watching him with familiar glances.

6

Understanding the Universe

Challenges and directions in modern observational astronomy

HARLAN J. SMITH

The task of astronomy is simple enough to state, namely *to understand the universe*.

Complete achievement of this goal is of course beyond any possibility of attainment. The impossibility arises in part because of the essentially infinite size of the task. In addition is the philosophical paradox of any infinitesimal part of the universe entirely comprehending the whole. Yet the astonishing thing is how far our insights have already reached, at least in broad-brush outline.

This chapter first considers different points of view from which one might understand the concept of understanding the universe, then briefly discusses some examples of where we might be going from here.

Discovery

Without objects to consider, astronomy would be a rather empty set. Indeed, in the popular mind the discovery of new objects is probably the most important aspect of astronomy. The finding of a planet beyond Pluto would bring headlines in most of the world's newspapers, though it would be of only minor importance in understanding the solar system and of no relevance to the vast remainder of astronomy. In a similar vein astronomers are occasionally asked whether they have discovered any new stars lately. Since just our own galaxy holds some hundreds of billions† of stars this is not difficult to do, but by itself offers singularly little insight.

Discovery of the different *kinds* of things which the universe contains is really the initial step toward understanding. The realization that planets exist, as substantial bodies orbiting a central sun, was the crucial step in beginning to understand the solar system. The discovery that stars are great luminous

† Throughout this chapter the American billion denotes a thousand million.

balls of gas – suns at great distances from us – was fundamental in the next step of exploration. Discovery of the galaxies, as vast separate assemblages of stars, gas and dust, opened up the larger universe to meaningful questioning. Discovery within the past quarter century of the existence of quasars carries the seeds of radically different physical understanding, also of limits to the observable universe.

How many significantly different kinds of things does the universe contain? Martin Harwit has attempted a rational answer to this question in his recent book *Cosmic Discovery*. While the definition of 'different' is slippery, most astronomers would probably accept his conclusion that to order of magnitude we have so far discovered about fifty different kinds of astronomical objects. He notes that really new things normally come into our ken when we explore new domains of observation, such as extended wavelength or energy of photons, polarization of light, increased precision of angular measurement or sensitivity of detector, etc. By noting the rate at which new kinds of things have been found as we have extended our senses in the past, and by delineating the theoretical limits which can ever be achieved in each of the possible dimensions of observability, he estimates that we have already discovered more than a tenth – perhaps even more than a third – of all the possibilities. And most of these discoveries have been made in the last few decades.

Taken superficially this plausible argument suggests that astronomy in one sense is a fundamentally limited science. With the still accelerating rate of discovery we can forsee the time several generations hence when virtually all the importantly different kinds of things the universe contains should have been discovered.

The notion that there may well be finite and exhaustable limits to this kind of astronomical discovery is depressing to some astronomers, albeit remarkable from the point of view of potential human achievement. Yet, even if this insight should prove correct, astronomy will not be out of business.

It is crucial to note that the discovery of a new kind of denizen of the universe is only the beginning of a process which in extreme cases may go on as long as the human race should pursue science. Studies inherently tend to move toward progressively greater depth and detail. From this point of view science has somewhat the character of certain Mandelbrot fractal sets – as each element is examined more closely, it displays its own rich branching structure rather similar to the larger units, and so on virtually ad infinitum.

As an example, the brighter planets were 'discovered' long before the dawn of recorded history, yet we are just beginning to get some crude first-order understanding of their real properties. Consider the hundreds of thousands of geographers and geologists, oceanographers and meteorologists, who have studied the earth, yet how they have so far literally only scratched the surface. Each of the planets and satellites in the solar system will need to be subjected to the same detailed kind of scrutiny before we will be able to understand most of what is knowable, much of it interesting and important, about the solar system. How much more will this prove to be true as our descendants explore the larger universe. In this sense I foresee that astronomy in the broadest sense will continue as one of the major human endeavours quite literally to the end of our time as a sentient race.

To be sure, as the rate of new discoveries falls off and the cost of making them increases, it is possible to imagine public support for cosmic search diminishing. Astronomers would find that unfortunate for several reasons. Consider simply the question of catalogs. If in principle something can be known, it is likely that someone will try to learn it. Humanity has much of this instinct. As extreme but nevertheless plausible examples I confidently expect that, when it becomes technically possible to do so, efforts will be made to identify and record the properties of all the trillions of comets believed to be in the great Oort cloud ellipting the sun in stately distant orbits, or to catalog and detail the properties of all the stars in our galaxy or of all the galaxies in the observable universe. Such activities might seem an exercise in futility and boredom. Yet even if nothing else were involved there is a kind of security in knowing what's out there, in reducing the level of potential surprise.

More to the point, the effort of finding, classifying and cataloging more objects of any given kind is important scientifically. Only in this way do the statistical regularities appear from which it is possible to draw often deep insights. From such studies unexpected anomalies can appear, leading to basically new discoveries within a field already partly explored. While to be sure there are severe factors of diminishing returns, it is likely that our descendants – controlling stupendous resources of energy, materials and time – will continue to explore and catalog their environments out to the limits of feasibility, and these limits will continue to expand with time. Since the known universe contains an indefinitely large number of major discrete objects, this is one sense in which astronomy has no practical limit.

Insight

The above arguments point up the conclusion that human knowledge of the universe will be forever incomplete. But our goals of course go far beyond discovering or cataloging its contents. True understanding of the universe requires insight into its phenomena, laws and course of evolution.

The objects populating the universe are far from passive. Most of them show a wide range of interesting phenomena – changes over various time scales. Yet, whereas the number of objects is effectively nearly infinite, the number of really discrete phenomena is rather small, probably only a few orders of magnitude greater than the number of kinds of objects in the universe. Thus the asteroids experienced initial creation processes, and continue to undergo a few slow orbital, collisional, rotational, erosional and similar processes which we must fathom if we are to understand the class of asteroids. With stars the range of phenomena is far greater, but still quite limited. It is possible to imagine that, on the same sort of time scale as Harwit's program of cosmic discovery, the human race will note and come to understand essentially the full suite of phenomena displayed by all the different kinds of inanimate objects in the universe. Again, if this should prove a true prediction, it could take much of the fun out of the pursuit of astronomy. One saving factor may be that, the closer we get to the end, the harder it will be to make asymptotic progress toward the goal of completeness. Since humans tend to perceive on logarithmic scales, overcoming each constant fraction of the residual unknown may seem as challenging and rewarding as the much larger absolute jumps made in the past.

On a higher plane of abstraction, all the objects and phenomena in the universe are presumed to be governed by the same set of natural laws. Astronomy has played a substantial role in their elucidation. Gravitation, mechanics, optics, electromagnetic theory, nuclear physics, relativity – these and other parts of physics stemmed in part from astronomical studies. This process is continuing. Astronomical research may be the only way in which several of the most fundamental questions of physics can be tested. These bear on whether there is only a single underlying physical theory, presumably of sublime clarity and simplicity, from which all the other laws and phenomena can in principle be derived. Most physicists hold this as the ultimate goal, and progress over the past several hundred years has strongly encouraged belief in it.

Yet, if and when this grail has been achieved it will not be the end of the endeavour to comprehend the universe. Understanding of objects and phenomena will still be very difficult in many cases because of the sheer complexity of the consequences of what appear to be simple laws. A classic example from ordinary Newtonian gravity is the ease of complete solution of the two-body problem, but the insolubility (except by numerical approximations) of the general three-body case. When electromagnetic fields or relativity theories are added the complexities exponentiate.

From still another point of view, physical laws describe what may happen – but they do not say just what did or will happen. Eddington, for example, pointed out the clear distinction between the general laws of solar system formation which could be derived from a sufficiently good understanding of physical laws, and the unpredictable special facts such as the presence of the specific planet Pluto in our solar system.

Likewise, the full story of the evolution, past and future, of the universe in which we find ourselves represents a challenge to human understanding which can never be completed. Great progress in sketching out the general picture has already been made. But even if virtually all else should finally be brought under strong control, complete certainty concerning the very beginning and the ultimate fate of the universe is likely to remain beyond our ken.

To summarize, several major threads run through modern observational astronomy:

> The discovery of new kinds of objects or phenomena, most often made serendipitously in the course of other work. This constitutes an exciting but rather transient phase of the science. Our single century – out of the thousands which have gone before and the millions which we hope will follow – is being privileged and challenged to make perhaps as many as half of all the possible discoveries of the fundamentally different kinds of things and processes which populate the universe.
>
> The drive to catalog, which has an unlimited future.
>
> The need for understanding. This absolutely requires the above kinds of data, but must be supplemented by physical theory which often requires astronomical studies for its corroboration, even for its genesis.
>
> The drive toward ultimate insight, whether or not attainable, as to the origin, evolution, and future of the cosmos.

Unfinished business

What are some specific challenges for modern astronomy, toward improving our understanding of the universe?

In the light of the four threads spun out above, the first type of challenge is discovery. Things which are fundamentally new tend by their very nature to be unsuspected. However, two effective strategies are normally open to astronomers who would like to make discoveries. They must keep an alert and open mind while going about their regular work, and where possible they should point their search toward observationally unexplored areas, this often requiring the development of new instrumentation. To be sure, predictions of theorists occasionally guide the search, as in the recent example of neutron stars. But even here the actual discovery took a very unexpected path through an accidental detection of radio emissions from pulsars.

Several key areas of observational and instrumental technique offer major prospects for new discoveries over the next several decades. Extending the wavelength range of observation of electromagnetic radiation is the most obvious.

For thousands of years the human eye was the only available detector of radiation. Although a truly amazing device, the eye has a factor-of-two limitation of wavelength sensitivity, from deep violet to dark red (4×10^{-5} to 7×10^{-5} cm). This ensured the effective invisibility of a great many objects and phenomena in the universe which don't happen to radiate strongly in the visible band.

Extension toward long wavelength came first with the discovery more than a century ago of the near-infrared. Such radiations were very gradually exploited for astronomical observation out to wavelengths about 25 times longer than visible light. Fifty years ago the much longer wavelengths of radio also began to be tapped for their astronomical information.

These long-wave spectral regions suffer blockage of some of their most important wavelengths. In particular, molecules in the earth's atmosphere afflict most of a wide band, by absorbing nearly everything from the shortest radio waves of about a millimeter on down to infrared waves of about 0.01 mm. Yet the cooler objects in the universe have their strongest radiations in precisely this band. This is because they emit primarily thermal radiation from their surfaces which tend to have temperatures in the range of around 3 to 300 Kelvin (absolute). Since objects or grains of dust far from some source of heat will tend to settle down in temperature toward the lower end of this range, it is clear that any census of the contents of the universe must be able to work with such radiation. Infrared and submillimeter telescopes located on very high dry mountain peaks, and airborne telescopes, can exploit at relatively low cost some partial gaps in the atmospheric absorptions. For the next ten or twenty years these will be among the most productive of all astronomical instruments. Nevertheless, far-infrared and submillimeter observations ultimately will be made almost entirely with orbiting or lunar-based telescopes.

For radio wavelengths above this molecular barrier, the earth's atmosphere becomes again transparent, opening another large window to the universe. However, free electrons in the terrestrial ionosphere effectively block radio astronomy observations at wavelengths much longer than about 30 meters. This is unfortunate, because waves ten to a hundred times longer carry much information about the extended plasma clouds of charged particles in and around the galaxies in space. Accordingly an observatory to explore this very long wavelength range will probably be built on the moon within the next two or three decades. Though in principle even longer waves would be interesting to study, interplanetary and interstellar scattering and absorption will largely preclude their use. But by the time such limits are reached, these various extensions to our observational power will span nearly a billion-fold increase in accessible wavelengths beyond the old red limit of the astronomer's eye.

The other direction, toward wavelengths shorter than blue, began to be opened a crack around the turn of this century, but it was soon discovered that an ozone layer high in the earth's atmosphere totally blocks ultraviolet radiation shorter than about 3×10^{-5} cm. So the advent of deep ultraviolet, X-ray and gamma-ray astronomy had to await the space age. Extremely important pioneering work has now been done in each of these wavelength regions, where the photons are so energetic that they are usually specified in units of eV (electron-volts of energy)†, rather than in equivalent units of wavelength.

The biggest breakthroughs in short-wavelength astronomy are surely still to come, over a very long future, as ever more powerful true observatories are orbited for longer periods of time. The first of these great space observatories is the Hubble Space Telescope, with its strong penetration into the ultraviolet. Analogous space observatories with very long lifetimes are needed to work with soft imagable X-rays up to about 10^3–10^4 eV. Still harder X-rays and gamma rays cannot be reflected to form images, but appropriate arrays of detectors in space can study them effectively out to indefinitely high limits which may ultimately exceed 10^{20} eV (although ground-based systems studying the cascade of atmospheric particles produced when such particles impact the upper air may be more effective for the most extreme energies).

The problem in carrying out this program lies in its great expense. The Hubble telescope alone will have substantially exceeded a cost of $2 billion within its first decade of use. International cooperation in paying for these splendid facilities must be part of the answer. There will also be a substantial drop in cost of space instruments, as progress is made in space technology and as large-scale orbiting and lunar stations are established. The results of putting short-wave astronomy more fully into space will be worth the effort. This extension will permit studying the universe in detail with photons reaching down in wavelength to a billion billion times shorter than those of visible light. In turn such photons are the products of the most violent objects and processes in the universe. These include neutron stars, black holes, quasars, gamma-ray bursters, ... singular end-points, and in some cases perhaps singular origins, of truly remarkable bodies and events.

Great observational progress is possible in other ways than wavelength extension. For example, angular resolution is a vital characteristic at every wavelength of observation. It sets the limit on the amount of spatial detail which can be observed. From the ground, optical telescopes have generally been restricted to only a little better than second-of-arc resolution, in large part because of the distorting effects of the earth's atmosphere. Important improvements down to a tenth of an arcsecond resolution will be possible in

---

† 1 eV corresponds to about 12×10^{-5} cm. As photon energy increases, the associated wavelength decreases.

the near future by building better telescopes at atmospherically superior locations, and using them at relatively long infrared wavelengths with devices able to compensate in part for atmospheric wavefront distortions ('adaptive optics'). Various forms of interferometry will give still further partial resolution gains exceeding the hundredth of an arcsecond level.

Again, however, it will be necessary to go to space to take full advantage of the possibilities for high resolution. Orbiting systems are already being designed which will approach the micro-arcsecond level. Nor does this even come close to the ultimate limits of resolution. These will be set, as in radio astronomy, by the distorting effects of interstellar gas clouds. But in principle it should some day be possible in space to build gigantic optical/infrared telescopes able to map at least the coarse details on the surfaces of planets around nearby stars out to distances of many light years – planets a million times more distant than those in our own solar system.

Interferometry involves combining the signals from separated apertures in such a way as to gain much of the resolution which would be possessed by a telescope actually as big as the separation. Radio astronomers using the technique of very long baseline interferometry have already achieved resolutions of ten-thousandths of an arcsecond on tiny bright objects, especially the nuclei of active galaxies. Since their present baselines are a large fraction of the diameter of the earth, progress here will depend on going to shorter wavelengths, gaining perhaps another factor of ten, and on going into space with one or more of the antennas – offering another order of magnitude improvement in the near future, and much more at later dates.

Major advances are also afoot in the methods by which light is detected and analyzed. Within the past decade, solid-state chips have proved able to improve over photography by more than 30-fold in the efficiency of detection of light. With modest further gains in their quantum efficiency and reductions in their noise levels, along with their production in larger sizes and range of wavelength-sensitivity, these devices will allow astronomers to approach a nirvana scarcely dreamed of by earlier generations – namely the ability to count and know something about the origin of nearly every photon so laboriously and expensively collected by the telescope. Such developments put a crushing burden on the data storage and analysis facilities required, but the extraordinary progress being made by the computer industry suggests that this problem can also be solved.

The intrinsic characteristics of the photons are also of highest importance. Measurements of the relative numbers of photons at each tiny interval of wavelength provide the raw material for the science of spectroscopy, by far the most insightful of all the subdivisions of astronomy. Analysis of spectra gives an extraordinary amount of detailed data concerning sources of light.

Such information includes atomic and molecular composition and abundance, temperature and pressure, rotation, velocity in the line of sight, and magnetic field amplitude and direction. In turn, from these data are derived important properties of the source such as distance, age, and state of evolution.

Spectroscopy is already such a high art that little further improvement in spectral resolution is likely, or perhaps even useful. But improvements in precision of determining wavelengths, currently in progress, offer the possibility of observing on stellar surfaces subtle waves having speeds of less than a meter per second. These waves, analogous to those produced by earthquakes on earth, probe the deep interiors of the stars, offering for the first time a way of looking beneath their surfaces.

Until recently, essentially the only properties of light utilized by astronomers were the wavelength, intensity, and direction from which it came. Polarization and phase information are now beginning to be used to a significant degree. A new science, quantum optics, may offer growing insights into what can be deduced from meticulous inquiry into higher-order effects of polarization, coherence, and correlation.

A number of exotic technical developments have been noted in this brief review. Important as they will become, the heart of observational astronomy for many years will continue to lie in what can be learned from photons collected by ground-based telescopes in the accessible ultraviolet to infrared range. These photons come from sources having temperatures in the range from hundreds of degrees up to hundreds of thousands of degrees – a substantial sampler of matter in the universe. The rich arrays of atomic and molecular transitions radiating or absorbing light in this wavelength range, the extraordinary amount of information which can be deduced from them, and the effectively unlimited number of fascinating objects to study by their means, guarantee that groundbased optical/IR telescopes will continue to be mainstays of astronomy for a long time to come.

It is nevertheless true that substantial advantages would accrue if nearly all astronomical observing activity could be carried out above the earth's atmosphere. However, at present it costs many hundreds of times as much to orbit a given collecting aperture than to build and operate it on the ground. Despite its enormous expense the Hubble Space Telescope is only a modest 2.4 meters in size. Since photons can be analyzed spectroscopically just as well on the ground as in space, most of this primary branch of astronomy will continue to utilize the relatively low-cost groundbased telescopes for several generations to come. Indeed, the greatest wave of telescope-building in history is just beginning. Whereas the largest existing telescopes are the dozen or so in the 4 to 6-meter class, a nearly equal number are now being planned in the 8 to 15-meter class. By the end of the century this should lead to nearly a factor of ten increase in the world's ability to collect and analyze light from the universe.

Until now virtually all our information about the universe has been received in the form of the various wavelengths of 'light' so far discussed. But physicists are learning how to make detectors for other constituents, which probably form a larger part of the universe than even the matter we can detect by its electromagnetic radiations. These constituents are of at least three distinct kinds: neutrinos, unknown weakly interacting massive particles, and gravitational waves.

Although the cross-section for interaction of neutrinos with matter is exceedingly small, the sun is a copious emitter. The flood of neutrinos pouring past and through the earth is sufficient to induce a few atoms of chlorine a day, in a swimming-pool-size tank, to change into argon atoms which can then be counted. Despite the difficulties of the experiment, it is now clear that the detected flux of neutrinos is only about a third of that predicted by what had been believed to be a rather secure theory of the generation of thermonuclear energy in the solar core. This anomaly – one of the most severe in modern astronomy – is being attacked on many fronts. One of the most promising will be the construction within the next decade of neutrino detectors using gallium, sensitive to a range of neutrino energy more directly diagnostic of the solar reactions. Efforts are also under way to instrument deep lakes and even cubic kilometers of the deep ocean with scintillation detectors in order to try to detect neutrinos in energy ranges which should be emitted by supernova explosions and perhaps other super-energetic celestial sources.

A second major dissonance between theory and observation in modern astronomy concerns several observational and theoretical evidences that the great preponderance of the mass in the universe is in a form or forms that astronomers cannot directly observe, at least in terms of the familiar baryons and leptons making up ordinary matter. Grand unified theories (GUTs) of elementary particles suggest the existence of very massive, very weakly interacting entities given such endearing names as axions, photinos, winos, wimps, etc. Recently it has been proposed that even some of these particles may be detectable by their effects on tiny supercooled domains in magnetic fields.

Finally, by far the most sensitive instruments ever conceived are now being built in a number of places to try to detect the gravitational waves predicted to exist by Einstein's General Theory of Relativity. Such waves would carry energy away from certain very close stellar binary systems. Effects arising from this mode of energy loss have almost certainly been detected by radio astronomers in a binary pulsar. In the 1970s direct detection of gravitational waves was claimed from the use of special detectors in two widely separated terrestrial laboratories. Later work has failed to confirm these observations, down to far greater levels of sensitivity. But new systems soon to come on line may push the detection threshold even below the energy of single quantum fluctuations. Success with these 'gravity-wave telescopes' will open up an entire new domain of astronomy, giving the possibility of direct observations of such extreme events as the collapse of stellar cores, the orbiting or even collision of neutron stars, the interactions of major masses with black holes, perhaps gravity waves still echoing across the universe since the violence of the big-bang origin.

With such large and almost totally unexplored areas of observation still to be opened up, it is virtually certain that many radically new and exciting discoveries will be made. Much of the work being done by today's astronomers was completely undreamed of fifty years ago. This will surely be the case for our children as well.

Insight

What major challenges to our understanding will be tackled by today's and tomorrow's astronomers with the aid of these magnificent tools? This survey can touch on only a few examples from some of the major research areas.

Nearest at hand is the solar system, the oldest realm of astronomical investigation. It has long been regarded as passé by many astronomers. Yet over the last few years it has become a more lively branch of research than ever before. Questions can now be meaningfully asked which could only be speculated on in the past.

Consider the closest bodies of all – the Earth–Moon system. After centuries of research and even after manned expeditions to the Moon, we still do not know how it was formed. Theories that the Moon originated as the smaller twin sister, the natural daughter, or the adopted daughter of the Earth continue to find defenders, although there are strong arguments against all of these scenarios. Evidence is gathering that there may have been a catastrophic collision of a large protoplanet with the primitive Earth which not only shattered the intruder but also blasted out a considerable amount of terrestrial material. Some of these fragments are imagined to have been captured into orbits around the Earth, and to have gradually coalesced through further gentler collisions into the Moon as we see it today. But was this really the story?

The sun is the nearest star. The mystery of the solar neutrinos has already been touched on. Among many ideas proposed to account for this grave anomaly are several that deal with the character of the deeply hidden core from which the energy emanates and which contains most of the solar mass. Does it hold concentrations of one or more of the mysterious weakly interacting particles? Is it perhaps rotating much faster than the outer layers, long ago slowed by magnetic drag against ionized gases in the Solar System? Has the central nuclear furnace been temporarily banked because of some geologically slow pulsation process? Answers to these and other hypotheses will come in part from detailed soundings of the deep solar interior now being undertaken by astronomers who are mapping the vertical speeds of tiny portions of the solar surface with errors as small as 10 centimeters per second. By analyzing the frequencies and wavelengths present in hundreds of hours of such data from many points on the solar surface, they detect waves which probe great depths into the sun. In turn, only a certain small range of models of the solar interior is consistent with the character of the detected waves. In this way, a combination of observation, calculation and theory promises understanding as to the real character of what, until a few years ago, was regarded as a region of the universe so very nearby yet quite hidden from view.

These solar studies may have important bearing on human welfare. Fluctuations of temperature and rainfall affect the prosperity, even the survival, of all who depend on the land. The climatic record suggests possible correlations with short-term (decades to centuries) solar-cycle variations, but more research is needed to establish whether these correlations are real, and if so what may be the physical links between solar magnetic field changes and their terrestrial consequences.

On the long time scale, it is still quite unclear what causes the great episodic ice ages, lasting for millions of years, although here again the sun may play an important role. But within the current ice age, intermediate-scale fluctuations having durations typically of many tens of thousands of years have been shown to correlate strongly with earth orbital changes, as proposed more than fifty years ago by Milankovich.

Geological factors such as volcanism and continental drift are probably also very important to climate change. But human intervention is now beginning to play a role as well. Destruction of tropical forests releases previously fixed carbon into the atmosphere, and reduces the ability of the earth as a whole to recapture it from the air. Industrialization takes fossil fuels from the ground and likewise converts much of them into atmospheric carbon dioxide. Even small amounts of carbon dioxide act as a very effective blanket, diminishing the ability of the earth to radiate its heat to space. Astronomers discovered this problem in studying Venus, whose carbon dioxide atmosphere leads to the surface temperature of 480 °C which only Dante might find congenial. Over the next few decades the terrestrial effects of enhanced carbon dioxide supplemented by other released gases such as freons are likely to become apparent in the form of a systematic warming of the earth by several degrees. Although the consequences are still quite uncertain, this may cause poleward expansion of the desert regions flanking the equator, and significant rises of sea level resulting from the melting of polar ice.

Astronomers have recently made the world aware of an inverse effect, that of a catastrophic man-made 'winter', which might result from climatic consequences of a large-scale nuclear war. This work was triggered by discovery of major coolings of Mars which in turn stem from periodic planet-wide dust storms. A major challenge for astronomy, working with other relevant disciplines, will be to understand both short and long term climatic drivers not only from the abstract scientific point of view but also in sufficient detail to be of practical use to the policy makers who must make political and economic decisions.

The smaller bodies of the solar system, the comets and asteroids, are also presenting exciting new challenges to astronomers for practical as well as scientific reasons. A picture is beginning to emerge as to their origin. The asteroids apparently represent fragments of a population of rocky, somewhat iron-rich, protoplanets formed at the time the terrestrial planets were also condensing from the hot inner regions of the proto-solar nebula. The comets were probably similar objects formed farther out in the solar system where temperatures were low enough to let the abundant volatiles, mainly of water and carbon dioxide, freeze and agglomerate into objects up to tens of kilometers in diameter. Near encounters with the massive outer planets subsequently nudged or hurled nearly all of these ice-balls far out into space. Those not given quite enough energy to escape the sun's pull gradually assembled into the present Oort Cloud consisting of vast numbers of comets orbiting mostly at great distances around the solar system. However, the detailed compositions and characteristics of these small bodies, and the specific ways in which they acquired their present locations in the solar system, are subjects of intense current research.

As is the case with solar astronomy, studies of the minor solar system bodies are of interest not only to astronomers but also to the human race as a whole. Such objects represent potential raw material for large-scale space utilization. Lifting great masses out to space against the earth's powerful gravitational field will remain expensive for some time to come. Comets and asteroids constitute an almost inexhaustible reservoir of material already there, rich in the key volatile elements hydrogen, carbon and nitrogen which are grossly deficient in the other nearby convenient space source, the Moon. Discovering and cataloging the properties of these objects is becoming an astronomical growth industry. Over the next half century the fruits of this work will likely be a giant economic enterprise of space industrialization.

There is yet a third reason why these minor bodies represent an important astronomical challenge. Evidence is growing that wholesale extinctions of terrestrial life forms are caused every few tens of millions of years by impacts of just such asteroids or comets, and that merely greatly-devastating impacts are far more common still. The odds remain rather good that we shall get through next year without suffering such an encounter. Nevertheless the results could be so catastrophic, with deaths in the hundreds of millions of people and costs in the tens of trillions of dollars, that it would seem reasonable for the human race to pay the astronomers' price of insurance. This would consist of the resources needed to discover and determine the orbits of essentially all such objects which could represent a hazard. Such a celestial 'coast-guard iceberg patrol' should also be charged with visiting and revising the orbits of any prospective candidates for collision over the next few millenia. Orbital revision of this kind should be straightforward for our engineering grandchildren. It has even been suggested that here may be a first truly constructive use of nuclear explosives.

Consider now solar systems on the larger scale. One of the key questions of modern astronomy is the exact way in which such structures are born. The past forty years have seen great progress. We no longer take seriously the baroque and exceedingly improbable stellar-collisional models which were in vogue for several earlier generations. We now know that stars form from great clouds of interstellar matter. These clouds, even when extended over light years in size, have some spin. As they condense under their own self-gravity in the process of star formation, the conservation of angular momentum practically ensures the formation of a great rotating nebular disk of gas and dust, from which protoplanets may grow.

This picture is appealing and probably generally correct. But the gaps in our understanding are great. To mention a few: Can we be sure what finally triggers the collapse of an interstellar cloud which has been placidly enduring its tenuous existence for hundreds of millions of years? A nearby supernova explosion? A spiral density shock wave of some kind? Stochastic collisions of smaller clouds? Any or all of the above, and more?

Once the collapse into a protostellar nebula occurs, how does the system get rid of the excess angular momentum which would normally make it nearly impossible for a central star to form at all? What are the roles of binary star formation in this process, versus the effects of magnetic field drag passing angular momentum from inner, more rapidly rotating regions toward the outer fringes of the cloud? Indeed, what is the true frequency of binary stars as opposed to the apparently rather uncommon single stars in space? Why do stars in the process of being born appear so often to show great activity, and is this activity essential to clearing out the placental materials from which protoplanets may have been growing? What conditions are necessary for true planets to form in orbit around a newly born star? What is the frequency of true planetary systems? Not a single planet outside our own solar system is yet known, although infrared astronomers are searching for binary systems containing low-mass 'brown-dwarf' companion stars which shine by virtue of their contraction, and are finding clear evidence of disks of dust around some nearby stars.

The evolution of stars, once they are formed, is now fairly well understood both in broad outline and some detail. This has been one of the great success stories of the past twenty years, stemming from an unusually fruitful combination of excellent observations, insightful theory, and powerful new computers. A few specialists are even going so far as to suggest that computers may become the most important tools of astronomy, with theory serving to guide the input and observation only needed to check the output.

Whatever their ultimate role, in this instance computers have been used to generate theoretical models of stars with an appropriate range of such input parameters as initial composition, mass, thermonuclear generation rates, radiative opacities, convective heat transfer, etc. The computer follows the internal and external changes which must take place as the matter evolves through its phase of initial heating under contraction to the point where, given enough initial mass, the star ignites the hydrogen-to-helium thermonuclear fusion furnace in the core. This places the star on the so-called main sequence, where it will spend much of its lifetime. Fusion gradually changes the chemical composition of the core, hence the structure of the star. Depending primarily on the mass, there arises a varied suite of swellings and shrinkings, surface heatings and coolings, over time scales measured typically in the millions to billions of years but punctuated by episodes during which the star also displays rapid pulsation.

It is remarkable how well these calculations reproduce the kinds of stars which are actually found in the universe. They give confidence that we have achieved real insight into the physical nature of stars. But now the experts are becoming interested in details, such as more accurate fits of models and observations, alternate ways of producing specific results, and the effects of complications such as magnetic fields, rotation and binary companions which are omnipresent in the real universe. It will yet be a long time before, to use the words of Sir Arthur Eddington, we fully 'understand so simple a thing as a star'.

Another major thrust of modern astronomy is the goal of understanding the deaths of stars. Here also the general outline has become clear. Stars having less than a tenth of the solar mass never ignite their thermonuclear furnaces, yet have so much energy of contraction as to allow them to die simply by radiating feebly as brown dwarfs over times longer than the age of the universe. More massive objects, up to a few times the solar mass, finally exhaust the hydrogen fuel available in their cores. After one or two relatively brief excursions to becoming red giants they settle down as placid white dwarfs about the size of the earth, and again go on cooling for spans of tens of billions of years. Stars in a comparatively narrow range of masses above this find that as they attempt to become white dwarfs the pressure contributed by electrons is insufficient to sustain them against collapse under their own intense self-gravity. They shrink into neutron stars with a million times the mass of the earth packed into a tiny ball only a few kilometers in diameter. Finally, most stars of still greater mass probably have no alternative, as they near the end of their evolutionary rope, but to self-gravitationally implode on their cores, then explode as supernovae and/or collapse with at least their remnant mass into a neutron star or a black hole.

Once again, such a broadbrush picture conceals much ignorance of the details even of single and uncomplicated stars. Among the leading current questions are the actual masses which divide the stars headed toward such radically different fates, the detailed physics of the interiors of white dwarfs and especially of neutron stars, the specific reactions involved in supernova explosions, and the upper limit of mass which can actually form a star without self-destructing as it aggregates. Again, the addition of factors such as rotational flattening and differential rotation with magnetic fields greatly complicate the picture in some of these cases.

But the most interesting complications arise in the ubiquitous cases of binary stars. The fact that all hydrogen-burning stars swell up by hundreds of times as they become red giants means that any sufficiently nearby companion can capture much of the material of the red giant, thereby putting both stars on altered, even radically different evolutionary trajectories. And the process can be reversed, so that matter may be passed back and forth on several occasions. Especially when the recipient is a white dwarf, a neutron star or a black hole, the infalling matter forms an in-spiralling accretion disk which can be so bright as to outshine the stars in the system, or can display extraordinary properties such as violent bipolar flows and X-ray or gamma-ray bursts. Such accreted material can even be the indirect cause leading to nova explosions. The ranges of possible phenomena are so wide and complex, and the difficulties of observation are so great, that the study of such systems will no doubt happily occupy astronomers for the forseeable future.

Interstellar matter has been mentioned as raw material for stars. It is also debris from old ones. This is proving one of the most fruitful fields of astronomy. Among its many current problems are the identification of some of the major constituents of the interstellar medium, and the understanding of their solid-state physics, their origins, and their true distribution in and between our and other galaxies. A few of the interstellar particles – the cosmic rays – are endowed with extremely high energy. Many different origins have been suggested, since their discovery early in this century. Some of the activity associated with neutron stars looks plausible, but as yet there is no generally accepted detailed picture to account for the sources of cosmic rays.

Most beautiful of astronomical objects are the great galaxies, shining by the combined light of thousands of millions of stars, and so large that light needs tens or even hundreds of thousands of years to cross their spans. Although their true nature as 'island universes' of stars was suspected nearly 250 years ago, solid confirmation was only achieved in the mid-1920s. Since then astronomers have been struggling – usually with telescopes of insufficient power – to classify, interpret the structure, map the distributions, and especially find ways of determining accurate distances to galaxies. Even today, after several generations of intensive work, there is still disagreement by nearly a factor of two in the fundamental scales used by different groups of astronomers to assign distances to the more remote galaxies.

While these traditional problems continue to occupy a great deal of attention, newer studies include the dynamics of individual and colliding galaxies, the distribution of chemical elements within them, and their evolution over cosmic time scales. In particular the question of the 'missing mass' looms as a ghost over our understanding of galaxies. The rotations of galaxies and the gravitational pulls which they exert on each other testify to the presence of as much as an order of magnitude more mass than we can account for in the familiar forms of stars, gas and dust.

Evidence is now overwhelming that many and perhaps most galaxies possess an extremely massive black hole at their core. Such core black holes may range from as little as a few hundred up to tens of billions of solar masses. According to one plausible picture, low-mass cores produce radio galaxies, so called because they have anomalously large radio emission in proportion to their optical luminosity. More massive cores are associated with galaxies displaying extremely active nuclei, the Seyfert galaxies. On this picture, galaxies with gigantically massive cores, generating the greatest luminosities in the universe and spewing matter out to distances of millions of light years, are the famous quasars.

One generally accepted mechanism is currently able to account for the staggeringly large energy release of quasars from a region of space only a few light-hours or light-days in size. This is the transformation of gravitational potential energy into kinetic energy of infall, through an accretion disk spiralling into a super-massive black hole. Large black holes radiate essentially no matter or energy. Nevertheless they may be the most luminous objects in the universe, by virtue of converting a substantial fraction of infalling matter to energy ($E = mc^2$) before the matter crosses the event horizon and becomes swallowed up forever in the hole. In this model of quasars, their appetites are prodigious. To keep the more luminous ones shining they must be fed a more or less continuous diet of about a million earth-equivalents of mass per year!

The logistics of providing this gargantuan supply of mass offer something of a problem. Quasars appear to be up to hundreds of times as common during the early phase of the universe as they are today. One hypothesis is that they early-on exhausted most of the matter easily available for capture, and are now on starvation diets. In this case, even the relatively feeble radio galaxies and many quiescent galaxies as well may have monsters sleeping at their cores, waiting patiently for billions of years for a chance near-encounter with another galaxy to stir the remaining gas, dust and stars into orbits once again permitting the capture and swallowing of some of this matter by the hole, thereby partly and briefly reviving the brilliance of its youth.

The above picture is one of the most dramatic achievements in the history of astronomy, yet it is not certainly true. Several schools of thought take radical exception to this explanation of quasars, even to the reality of black holes. While these viewpoints now constitute rather tiny minorities, history warns us that new discoveries in astronomy and physics can overturn even very widely accepted paradigms. Accordingly one of the greatest challenges to astronomy is to test the present party line on black holes and their effects, and to struggle with the monumentally complex physics and mathematics required for their detailed analysis and interpretation.

As we scan the universe on ever-larger scales, it has become clear that most galaxies are organized in superclusters typically tens of millions of light years in size. Still more recently, evidence has emerged that, at least as far as bright galaxies are concerned, space consists mostly of gigantic voids, with the superclusters populating the edges. The universe thus may have a foamy characteristic, rather like a space filled with soap bubbles.

Cosmologists are currently struggling to understand whether the evolution of this structure was one wherein star clusters and galaxies formed first, later accreting into clusters, superclusters and voids, or whether initial conditions somehow produced the aboriginal matter in greatly and widely separated concentrations within which condensation into galaxies and stars later took place.

Observational astronomers are using present telescopes, while waiting impatiently for better ones, to probe this structure of the larger universe and, as the saying goes, to keep the theoreticians honest. So far the evidence is extremely strong that at least our present space-time and the matter which it contains arose in a Big Bang some ten to twenty billion years ago.

Four independent lines of investigation support the Big Bang cosmology, and give generally concurrent ages for the beginning:

The theory of stellar evolution coupled with observations of the most ancient stars so far detected indicates that they came into existence less than twenty billion years ago.

The ages of the oldest radioactive elements so far found seem to be in the range of ten or fifteen billion years.

The present rate of expansion of the universe, as measured from the red-shifts of the galaxies, is consistent with all the matter in the universe being together at a common origin some ten to twenty billion years ago.

Finally, a Big Bang origin would have flooded the universe with high-temperature radiation which by now would have cooled down to only about 3 degrees above absolute zero; such radiation is indeed observed.

Evidence for the Big Bang presents so solidly self-consistent a picture as to make it seem unlikely to be overturned. Yet here as in every other area of astronomy there are maverick opinions, and radically different explanations of the observed phenomena. And it would be surprising if – with questions so profound and on which we have had only a few decades to work – the human race really already has the answers.

If the consequences of the Big Bang in many ways do seem solid, the same cannot be said for the boldest theories of all – those of how the Big Bang itself began. Here science is discovered to be eating its tail. The investigations of the tiniest elementary particles prove to couple essentially and fundamentally with the theory of the universe as a whole. This is true both for the so-far earliest physically definable moment, an almost inconceivably brief 10^{-43} seconds after the beginning, and for the consequences down to the present. The 'missing mass' which astronomers seek in order to account for dynamical effects of galaxy rotations and interactions may well be in the still only-hypothecated weird and weakly-interacting particles produced during those vanishingly brief moments of creation of our universe. That mass may also serve to balance the universe between catastrophic expansion or collapse – again by an order of magnitude, ordinary matter appears to be insufficient to 'close the universe'.

Some physicists and cosmologists are rather optimistic that one day the human race will fully understand the creation of the universe, also its general future course to the end of time. They believe this will be so in terms of mathematical-physical concepts so general, so compelling, and so unique as to admit of no other possibilities. Einstein may have had something like this in mind when he said 'I wonder if God had any choice in creating the universe?'

But most astronomers, I believe, are skeptics. In a universe so inconceivably vast, will we ever really discover and understand *all* the relevant phenomena? Are there levels of 'reality' going deeper in the submicroscopic world than we will ever be able to penetrate with any tools or observations? Even the uttermost ultimately accessible slice of space and time which the human race can occupy is so infinitesimal on the scale of the universe as to give deep pause, both physically and metaphysically, as to whether we can ever comprehend it all.

Perhaps, if we are not alone as sentient beings, our remote descendants will find help from outside with these questions. Indeed, the search for company is likely to prove the most durable of all the challenges faced by astronomy, as long as the human race shall last. If no other intelligences are found, it will make the continued search for them ever more important. If they prove to exist, their variety will surely be infinitely interesting, and worthy of the most complete census which can be performed this side of eternity.

In turn, the deepest challenge, for our generation and the next several to follow, is to survive the lethal potentialities currently being unleashed by high technology on a world filled with people still inheriting and displaying stone-age reflexes. If this challenge is met, astronomy will not lack for challenges of its own.

7

Frontiers in cosmology

FRED HOYLE

My story concerns developments in cosmology that have taken place, or not taken place as the case may be, since the year 1965 when the so-called microwave background was explicitly detected by Penzias and Wilson, working at the New Jersey laboratories of the Bell Telephone Co.

Microwave radiation comprises wavelengths too short to be normally usable for radio purposes and too long to be considered as heat. The discovery of Penzias and Wilson was that radiation over the microwave range arrives at the Earth from space. It does so with remarkable uniformity from directions all over the sky. The radiation has closely the same characteristics that one would get from a body at a very low temperature, a temperature of about 2.7 K, for which reason the background is often referred to as the 2.7° radiation.

Of course there is no such actual body covering the whole of the sky. Where then, astronomers asked themselves, has the microwave radiation come from? The answer given was a remarkable one. Since nobody in 1965 could think of any explicit source for it, the argument was that the radiation exists because it existed before. It exists today because it existed yesterday, and it existed yesterday because it existed the day before that, and so on. Back to where? To the origin of the whole universe. So the radiation was not explained, any more than the universe itself was explained. It came into being *with* the universe, and it did so, not with its present day low temperature, but at an enormously *high* temperature, which has since been lowered progressively with time by what is known as the redshift effect, the redshift effect being an acknowledged consequence of the expansion of the universe, a phenomenon that is well-proven by astronomical observations and of which one can have no reasonable doubt. (At any rate so long as the word quasar is mercifully omitted.)

The possibility that the universe might have a microwave background had

97

been considered speculatively long before its actual discovery by Penzias and Wilson, considered notably by George Gamow and his associates. Gamow had sought to explain the origin of the chemical elements as a primordial consequence of a microwave background, as a relic of the early history of the universe when the background was extremely hot, the hot big-bang as it became called. This issue had become settled otherwise, however, already in the 1950s, so far as the great majority of the chemical elements are concerned, settled so far as all the elements heavier than lithium are concerned. The battle for the heavier elements had tended to obscure the situation for the very light elements, of which helium and the heavy isotope of hydrogen known as deuterium were the most important. These very light nuclei could still be relics of the early universe, and perhaps they might even have of necessity to be relics of an early universe.

Speaking personally, from 1963, still two years before the actual discovery of the microwave background, I set myself the task of repeating Gamow's calculations for deuterium and helium, using various improvements in physical theory and improvements also in the astronomical observations. Working at first with R.J. Tayler of the University of Sussex, and then with R.V. Wagoner and W.A. Fowler at the California Institute of Technology, it was shown that from a highly accurate knowledge of the primordial abundance ratio of helium to hydrogen, together with a more modestly accurate knowledge of the primordial ratio of deuterium to the common isotope of hydrogen, it was apparently possible to infer within the framework of big-bang cosmology a result of profound importance. Given such data, it was possible to infer the basic topological structure of the universe, whether the universe is open and infinite, or closed and finite. If the universe is open and infinite, it will go on expanding forever, with the galaxies separating everlasting from each other. One might say in such a case that after beginning with a big-bang the universe will end in a small-whimper. But if the universe is closed and finite, the present expansionary phase will be replaced by a contracting phase that ends in a reversed big-bang situation, with the universe eventually collapsing and going out of existence – just the reverse of its origin.

So what conclusion did we reach from the work of 1963–67? Subject to uncertainties in the astronomical data, and subject to the framework of big-bang cosmology, we decided tentatively that the universe is open and infinite, which is to say we human folk are *en route* from a bang to a whimper. Astronomers have worked hard in more recent years to improve the data, with results that tend to confirm the opinion of 1967. Tantalising uncertainties remain, however, especially about whether the deuterium and helium abundancies that have been thought to be primordial really are primordial. Evidence has come increasingly to light, about which I shall be speaking later, which suggests that our present generation of stars was preceded by an earlier generation of what have become known as Population III stars. Such Population III stars must certainly have produced some helium, perhaps a significant fraction, or even the whole, of what was previously thought to be primordial. And if cosmic rays were produced by pulsar-like stellar residues among the Population III stars, there would be collisions between cosmic rays and helium nuclei, inevitably producing some deuterium. How much deuterium we do not know. So what began as a seeming clear-cut argument is now tending to dissolve into uncertainty, as I fear happens in astronomy only too often – for the reason of course that our mental picture of the universe is a grossly oversimplified picture of the real universe.

There is more than one way of skinning the cat, of course! It is possible to decide whether the universe is open or closed in an entirely different way, by comparing the average density of matter in space (as it is presently observed) with what is referred to as the closure density. The closure density can be calculated unambiguously when the present-day rate of expansion of the galaxies has been determined from observation. The ambiguities arise, however, in the observations themselves, tantalisingly as always. Warring clans of astronomers differ about the expansion rate of the galaxies to within as much as a factor 2, which shows up as a factor 4 in the calculated value of the closure density, with estimates for the closure density ranging from about $5 \times 10^{-30} \, \mathrm{g\,cm^{-3}}$ on the low side to about $2 \times 10^{-29} \, \mathrm{g\,cm^{-3}}$ on the high side. Even the low value here is nevertheless quite a bit higher than the average smoothed-out density of stars in galaxies. To estimate this smoothed-out density it is necessary, first, to count galaxies so as to determine their present-day average density in space. This has to be done by observing moderately deeply into space. Then nearby galaxies are observed in detail to determine their stellar content. When all this has been done, an average density certainly not greater than $\frac{1}{10}$ of the closure density is obtained, which taken by itself would imply that the universe is quite markedly open, the same conclusion as was reached from the deuterium–helium argument of 1963–67.

But is most of the material of the present-day universe in the form of stars one next has to ask? What about gas, for example the very hot gas in clusters of galaxies that emits X-rays which can also be observed. Well, this gas turns out also to be too small to shift the argument significantly. Indeed, all forms of gas producing *observable* forms of radiation turn out to be significantly too small to shift the argument. Only gas that has so far been *unobservable* could be relevant, and this remaining possibility for gas has by now been pretty well whittled down to the possibility of there being comparatively cool ionized gas throughout intergalactic space, ionized hydrogen mostly. The disposition among astronomers, however, has always been to believe that what they do not observe themselves does not exist. So consensus opinion developed that the mean density of matter is indeed considerably less than the closure density, and the astronomical literature of ten years ago is full of statements to this effect. So it came to be believed, with belief elevated in some quarters almost to the status of fact, that the universe is open and infinite, that we are indeed journeying from big-bang to whimper.

But the real universe has a happy knack of trumping the scientist's trick, just at the moment when the game seems to him to be won. From developments in radioastronomy, and to a somewhat lesser extent in molecular astronomy, it became clear that the galaxies possess enormously massive halos, of the order of ten times larger in both dimension and mass than the visible stars. Thus the visible stars occupy only a small inner volume and they contribute only a small fraction to the masses of galaxies. The average density of matter in space has therefore to be lifted and it has to be done by material in a form that is not directly visible, which consideration leads to what has become known as the problem of the 'missing mass'. The concept of missing mass had the effect of moderating previous beliefs, so that many scientists began to wonder whether the average density of matter in space might not be comparable to the closure density after all.

What was this missing mass now? If it were ordinary matter, existing in some non-visible form – examples might be stellar residues (white dwarfs, neutron stars, black holes, planets) – then the observed deuterium could not be primordial, and one of the corner stones on which belief in big-bang cosmology had been built would collapse. There was an alternative, however, which did not lead to this conclusion. The missing mass might be in the form of neutrinos of a kind which possess an inherent small mass of their own, a so-called rest mass. Since neutrinos fitted readily into big-bang cosmology as survivors like the microwave background from the earliest moments of the universe, the consensus readily adopted itself to the view that one of the several known forms of neutrino must possess an inherent small rest mass. In this way big-bang cosmology would be saved, although the previous conclusion concerning the open topology of the universe might come under threat. And when experiments reported from the U.S.S.R. and from California claimed to have detected just such an inherent neutrino rest mass, the position to many astronomers seemed to be set fair again. But experiments were then reported by a trinational group of Swiss, German and American scientists working at Grenoble in France, which showed a negative result, and since the Grenoble experiment was clearly the most sensitive of the three, it is there that the issue embarrassingly now stands. Besides this, there are sound reasons for doubting that neutrinos with non-zero rest mass could solve the very problem they were supposed to solve, the problem of massive extended galactic halos. There could be such neutrinos in intergalactic space, or perhaps aggregated inside clusters of galaxies, but not aggregated around individual galaxies. So quite apart from the experimental situation, the idea fails just where it was needed.

There is an irony in all this, when you consider the furious antipathy of most scientists towards the religious creationists. The religious creationist says that because man's origin, or the origin of life if you like, cannot be explained to his satisfaction, life must have been created *ex nihilo*. The cosmologist says that because the microwave background cannot be explained to his satisfaction it must have appeared *ex nihilo* at the origin of the universe. Apart from the semantic difference that scientists use the word 'origin' instead of 'creation', the attitude of mind is the same. Indeed the scientific attitude was not new. It had existed for many years before the microwave background was discovered, and it came from a consensus assertion by physicists that the so-called baryonic component of matter could neither be created nor destroyed. If this were so, then the heavy component of matter could not have had an origin except at the origin of the universe, which is to say an event of creation *ex nihilo*, just the same situation as for the microwave background.

All this went sharply against my own 'faith' if you prefer a religious word, or against my 'gut feeling' if you prefer words more acceptable to the scientist. My predilection is for exactly the opposite state of affairs to the one I have described. I feel that every phenomenon we observe should be explicable in terms of scientific methodology. Everything must have a clear-cut origin or explanation of a scientific kind, describable within the mathematical structure of science. What one cannot discuss in my opinion is the origin of the universe itself. The universe is off-limits according to my point of view.

Because of my gut feeling in this respect, as long ago as 1948 I began to experiment with mathematical structures that permitted matter to originate at the expense of what I called a C-field, which in terms of particle physics would be carried by what is called a boson. Unwittingly, I made the semantic error of calling it a creation-field instead of an origin-field. I did this in innocent analogy to the term pair-creation which was, and still is, widespread in physics. Judge my astonishment on my first visit to the Soviet Union when I was told in all seriousness by Russian scientists that my ideas would have been more acceptable in Russia if a different form of words had been used. The words 'origin' or 'matter-forming' would be O.K., but 'creation' in the Soviet Union was definitely out.

One of the possible consequences of this early C-field cosmology was that the universe might be in a so-called steady state, with the origin of matter and the expansion of the universe in balance with each other. But the mathematical structure was actually much richer than that, with many other possibilities which J.V. Narlikar and I investigated partially in the 1960s.

I mention this C-field cosmology with its origin of matter, because by about 1980 particle physicists had decided their earlier consensus, that the baryon component of matter could neither be created nor destroyed, had been an error. With the immense skill which particle physicists always show in their public relations, they managed to represent this correction of a long-standing mistake as a great new discovery.

The way in which the correction of the mistake had come about was as follows. From the early 1960s the quark structure of the heavy component of matter had become well-established. Quarks are distinguished one from another by two distinct and separated properties, which were referred to as 'flavour' and as 'colour'. Notice again the penchant for semantics. By a clever choice of words the impression of an inner set which is clued-up as the Americans say, and an outer set which doesn't even know what the words mean is created. The words made the issue seem important, which is always a big advantage when it comes to obtaining large research grants from governments.

Interactions between ordinary baryonic particles, the well-known nuclear force, was seen to arise from connections between the 'colour' properties of the triplets of quarks composing the ordinary particles, interactions that took the mathematical form of three-dimensional unitary matrices, which had an analogy to the two-dimensional unitary matrices that were known already to describe the weak interaction force of β-decay, which had itself been success-fully joined to the electromagnetic force by the Weinberg–Salam theory. These several analogies and successes suggested the idea of unifying the three forces into what became known as a grand-unification-theory, GUT for short, another snappy piece of semantics you notice. Semantics apart, what was really involved was the submergence of three hitherto separated groups of mathematical transformations into the transformations of a single larger group. The situation hitherto was now seen as a partial representation of some larger group, a situation with which mathematicians are very familiar. The problem then for the physicist was to find the larger group, but unfor-tunately knowing the simpler form did not determine what the larger group must be, because there are many larger groups that could yield the simplified form in question.

I have, you will have noticed, had a bit of fun at the expense of the particle physicist and now I would like to redress the balance a little. Here we have an idea that is virtually certain to be correct, but without experiments designed to reveal its nature, the larger group cannot be found. The experi-ments cost money. A great deal of money unfortunately, to such a degree that there are now many people in Britain arguing that we in Britain should withdraw from this field of research, on the grounds that it is a field which can only become still more expensive, until ultimately, if given its head, it would bankrupt the entire human species. The first answer to this argument is that, although particle physics is costly, it is not costly to the point of bankrupting the whole human species. It is not as costly as space research. Particle physics will not benefit anybody in a practical way it is also argued. Argued falsely very likely, for who knows what will benefit the future? The crucial answer to these criticisms, which are being advanced in my country by an organisation whose activities consist very largely in funding second-class research, is that the issues of particle physics, like the one I have just described in detail, are issues for all time. Unlike the multitude of second-class projects which the British Government funds quite happily, and which collectively are far more expensive than particle physics, the issue here is timeless. So long as the human continues as an intelligent, technologically oriented animal, the truths of particle physics will long outlast those who make them, will long outlast the Pyramids of Egypt, and if in the far distant future we contrive as a species to leave this planet, the truths of particle physics will outlast the Sun itself. Any government which does not understand this, and which diverts finances from first-class research to activities that are quite obviously second-class, condemns its people in a very real and fundamental sense to a second-class future.

But I must desist from such impatient remarks and return to the frontiers of cosmology. In the absence of precise knowledge as to how to broaden the present mathematical description of the quark theory of matter, the natural thing to do is to choose the simplest possibility, where the quark interactions are rather naturally expressed by unitary matrices in five dimensions, the so-called GU5 group. The point to which I have been working is that even this simplest possible step contains new interrelations capable of changing the quark structure of matter, new in the sense of being additional to the old three forces – which is to say additional to the nuclear force, the electromagnetic force and the weak interaction. These new interrelations are carried by particles which have been given the name of X-bosons. Note again the clever semantics. Everybody knows that X stands for mystery, and so everybody wants to know what X-bosons are and what they do. One thing they can do is generate a field with the same crucial property as the C-field of twenty years ago, the property of possessing an energy density that is negative. And just as the C-field could give rise to the origin of the baryonic component of matter, so can the X-bosons. And just as the mathematical structure of C-field cosmology could give rise to what was called a steady-state condition, with the origin of matter balancing the expansion of the universe, so the origin of matter from X-bosons can give rise to a steady-state condition. Only of course the particle physicists use different semantics. Instead of a C-field they talk about 'false vacuum', and instead of the steady-state condition they talk about an 'inflationary phase'. So that people generally do not realise that it is just the same old theory, the theory which they formerly said was impossible, rearing its head again.

Because the GUT theory does not include gravitation on any new basis, protagonists of GUTs have regarded the universe as originating in much the same way as before. But even without explaining gravitation in terms of particle physics it is possible to show that quantum gravity as opposed to classical gravitation seems not to permit the universe to have an origin, as Narlikar and Padmanabhan have shown in a beautiful paper. Narlikar has therefore suggested that the basic structure of the universe is the old steady-state condition, or the inflationary condition if you prefer the semantics that way, but with a much larger choice for a critical coupling constant than we ever dreamt of using in the 1960s, a coupling which has at present to be chosen somewhat arbitrarily, because we do not know its numerical value either from physical experiments or astronomical observations. Otherwise the mathematical structure is similar to what it was before.

The next step takes the best from the two former theories of cosmology. Should the X-bosons fail to maintain the rate of origin of matter in a particular finite region of space, matter-creation eventually stops in that region, which then proceeds to expand like a big-bang cosmology. So you have big-bang situations appearing as bubbles in an otherwise maintained inflationary or steady-state universe. The bubbles expand more *slowly* than surrounding regions – they become bubbles because the origin of matter stops within them. Indeed the bubbles expand at just the rate which lies at the topological division between an open and a closed universe, just at closure as one says. Choosing the coupling constant heuristically from quantum considerations, the particular bubble in which we are living turns out to have had a size at the moment it began to form of only about one-tenth of a millimetre. The theory requires all the galaxies we see, the whole of the universe visible to astronomers, to have emerged from a region a mere $\frac{1}{10}$ of a millimetre in diameter, which gives you an exceedingly graphic idea of the immense particle densities which are contemplated nowadays in the GUTs.

Is the universe really like that, you will feel tempted to ask? I must admit that for myself I don't think so. If one looks at the details of the origin of matter, it is contemplated that matter and antimatter originate in equal amounts, which I have always thought to make for an impossible problem – especially at exceedingly high densities – in ever separating the two components into properly distinguishable aggregates. And the problem of the origin of the galaxies themselves remains obstinately without solution. One can contemplate two possibilities in tackling this intractable problem, one can try to argue that the galaxies formed before the so-called decoupling of matter and radiation, but in that case the masses which might conceivably condense are enormously too large. Or one can say that the galaxies formed after the decoupling of matter and radiation had taken place, when the opposite trouble arises – the condensations then have masses that are far too small. This problem has a long standing antiquity and it has never been solved with more plausibility than one associates with casual hand-waving. If one introduces a *deus ex machina* that somehow contrives to promote the formation of galaxies, the condensation process should show, still to this day, as a weak signature written on the microwave background. Such a signature has been looked for with great care and effort, but it has not been found. While the present status of the observations does not quite rule out every possible big-bang model, it does rule out many models, and it lies at the margin of significance for the rest of them.

What these observations appear to be telling us rather insistently is that the microwave background was generated at an epoch of the visible universe *after* the galaxies formed. If this turns out to be so, all the developments in cosmology that have taken place since the discovery of the microwave background in 1965 will stand in need of reassessment. But, for me, a still bigger point turns on a general matter of experience. Throughout the history of science, correct theories have always led to a stream of prolific discoveries. Yet I think most young students of cosmology today, if they were asked to explain their main reasons for believing in big-bang cosmology, would cite first the existence of the microwave background, and second the helium–deuterium argument I described earlier. In other words, the strongest support for big-bang cosmology would be seen to come from developments of the mid-1960s, not from the truly immense amount of research in cosmology which has taken place since then. Apart from the introduction of ideas from the GUTs, which I think may not be altogether welcome to big-bang supporters, the point is that 15 to 20 years of intense research has been largely arid. This does not seem to me the hallmark of a correct theory.

Where then do I personally think we should go from here? When around 1966 Narlikar and I first considered the idea of a bubble universe, I did my part of the investigation in the spirit that it is the job of the theoretician to investigate every possibility, rather than because I had a special affection for such a universe.

Today, however, I am more positively disposed towards the bubble universe, largely because of the possibility it offers of relating cosmology to the GUTs. Where I tend to differ from the discussion I have just given, is in the choice one makes for the so-far unknown coupling constant, the scaling factor that decides the initial size of the bubble in which we live. In effect, the scaling factor depends on the decay lifetime of the X-bosons responsible for the origin of the baryonic component of matter. Narlikar chose his decay lifetime from the well-known quantum result, correct in most cases, that lifetimes are of the order of Planck's constant divided by the energy involved, in this case the so-called Planck mass, which is a very large particle mass indeed, $\sqrt{(Gh/c^5)}$ where G is the gravitational constant. But there are examples in physics of lifetimes that are vastly longer than one would obtain in this way. One would obtain about 10^{-20} s for the decay of the neutron, whereas the actual lifetime is about 10^3 s, a discrepancy of order 10^{23}. For the nucleus ^{87}Rb, the discrepancy is greater still, by a factor of order 10^{39}.

One notices that such great extensions of the lifetime seems to occur when a higher symmetry breaks a lower symmetry, and the X-bosons are symmetry-breakers *par excellence*. They break the conservation of baryons, giving a proton-decay lifetime, not of 10^{-23} s, but in excess of 10^{31} years. The effect of an enormous increase in the lifetime of the X-bosons would be to make the initial size of the developing bubble which gave rise to our observable universe very much larger than the minute estimate of $\frac{1}{10}$ of a millimetre obtained by Narlikar. The scale could even lead to a situation more akin to the present-day observable universe than the theory considered by Narlikar and myself in 1966. What I have in mind is that the background of the inflationary phase of the universe involved an average density of matter such that interesting astrophysical processes could take place in that state, a density, say, of 10^{-20} g cm^{-3}. The situation would be rather like it is in the Giant Molecular Clouds of our present-day galaxy. But instead of being restricted as the present-day clouds are, such conditions would obtain on a vast scale. Stars could condense and produce variations that caused more and more stars to condense, thereby building what we know today as the microwave background. Somewhat similar ideas concerning the condensation of Population III stars have been considered by M.J. Rees and his colleagues, but not so far as I am aware as a source for the whole microwave background. Rees' theory was tied to standard big-bang cosmology, whereas the present ideas are tied to an inflationary universe related to the GUTs – or if one is obstinate about the semantics, to the old steady-state cosmology.

I have left what I believe to be the most significant point to the last, although if one accepts the claims of the biologists and biochemists of earlier decades, namely that life originated on the Earth, the point does not arise. But once one understands that the amazing superastronomical complexity of life could not have arisen according to what is quite frankly a ludicrous proposal, it becomes a part of cosmology to provide a background against which life might have developed. Cosmologies that do not meet this requirement will in my view turn out wrong. I must admit that the thought of a universe consisting everywhere of Giant Molecular Clouds, and with a background temperature of about 300 K – just right to provide warmth everywhere for living organisms – was a motive for me to consider the giant molecular cloud scenario I have just described, since of all cosmological situations it seems the one best equipped for coming to grips with the superastronomical complexity of life. But I must avoid claims that go too far, because experience in cosmology should have taught us that when we most think we are nearest to the truth we may be the furthest from it. The safest thing one can say about cosmology is that thirty years from now it is likely to be very different from what we think today. The key issue for cosmology, it seems to me, is the one I mentioned earlier, whether governments are intentionally going to desist from basic investigations of the physical laws that lie at the root of cosmology, and without which everything in cosmology, and in physics too I would think, will dissolve eventually into aimless speculation.

8

Did the Universe originate in a big bang?

(Some 'non-standard' views on cosmology)

JAYANT V. NARLIKAR

Modern cosmology is essentially a product of the twentieth century. On the observational side the subject was launched in 1929 by Edwin Hubble's discovery that the nebulae lying outside our Galaxy have redshifts that increase linearly with the nebular distances. Hubble's discovery was interpreted as a consequence of the expanding universe. The cause of expansion? A big explosion that caused particles of matter and radiation to move away from one another! The epoch of this explosion commonly called the 'big bang' is supposed to have marked the creation of the universe.

The notion of a big bang received observational support in 1965 when Arno Penzias and Robert Wilson found that the universe has a radiation background predominantly in the microwaves. With a presently estimated temperature of 3 K this background is believed to be the relic of the early post-big-bang era when the universe was extremely hot.

These observational results went hand in hand with theoretical developments. In 1915 Albert Einstein proposed his remarkable theory linking spacetime geometry to motion and gravity. Known as the general theory of relativity, this theory was used by him in 1917 to construct a simplified model of the large scale universe. Einstein's model was of a static universe and was therefore unable to account for Hubble's findings. However, Alexander Friedman in 1922 constructed expanding models that did anticipate Hubble's discovery.

The Friedman models imply that the universe has been expanding from a singular epoch in the past, an epoch when its space had zero volume and infinite curvature. The adjective 'singular' implies the breakdown of any sensible mathematical description. It was this epoch that later got the popular title of 'big bang'. If we associate a linear scale S with the universe, then in

Friedman's models the scale factor S has been increasing from its value zero at big bang. In the late 1940s, George Gamow argued that explosion implies a universe that was initially 'hot'. Gamow and his younger colleagues Ralph Alpher and Robert Herman tackled the physical problem of how such a hot universe played the role of a nuclear reactor for synthesizing atomic nuclei. One of their predictions was that there should be a relic radiation background in the microwaves surviving today.

It is interesting that both the major discoveries of 1929 and 1965 were anticipated theoretically. It is even more interesting that both theoretical predictions, of Friedman in 1922 and of Gamow *et al.* in 1950 were ignored at the time they were made. Certainly they played no role in inspiring the observations of Hubble and of Penzias and Wilson.

By hindsight the reason can be seen in the overall suspicion with which cosmology was viewed not only by physicists but even by astronomers. Cosmological theories were considered no more than speculative exercises without any possible testable status.

All that has changed dramatically in the last two decades! Not only is the big bang picture considered respectable, it is being used extensively by particle physicists, the high priests of science, as the testing ground for their programme of unification of physical theories. Working jointly, the particle physicists and cosmologists of today are far more daring than Gamow and his colleagues. While Gamow's discussion of the *early universe* during the first 1–200 seconds after the big bang was considered highly speculative four decades ago, the present work on the *very early universe* takes us to the era when the universe was only 10^{-37} second old!

Some reservations

While the cosmologists may rejoice at their subject being elevated to this high pedestal from the earlier status of a poor relation, viewed objectively this transformation has certain disturbing features. The marriage between cosmology and particle physics appears to be motivated by convenience rather than by appreciation of each other's virtues and weaknesses.

For, take the particle physicist's motive. His holy grail is the unification of all physical interactions. Einstein himself had felt deeply that there should be a single unified theory, although his lifelong attempts to deliver one did not succeed. Moreover, while Einstein during his life was in a small minority of physicists who believed in unification, today the situation is dramatically altered. The successful unification of electromagnetic and weak interactions has generated big momentum towards theories of 'grand' unification (GUTs) and 'super' symmetries (SUSY) in which eventually all laws of physics would be brought together.†

Unfortunately, the particle energies at which these theories can be dynamically tested are far higher (10^{15}–10^{17} GeV) than those achieved in manmade accelerators ($\sim 10^3$ GeV). Indeed this energy gap is too wide to be bridged by any foreseeable future technology. But then, theories which cannot ever be tested are no more than speculations, as any physicist will agree. Are GUTs and SUSY, with all their intellectual appeal, mere speculations?

The answer would have been 'yes' but for the big bang universe! It is here and only here that a brief era occurred in the past when the particle energies were so high that GUTs and SUSY had decisive roles to play. These theories can therefore be tested provided it is established that such a laboratory of the early universe ever existed. Hence the particle physicist has no alternative but to take the validity of the big bang for granted.

The arrival of the particle physicist on the scene has been timely from the cosmologist's point of view also. For, there are several questions concerned with the big bang cosmology that cannot be answered without inputs from particle physics. Why does the universe appear to contain predominantly matter rather than antimatter? Why are there so many (10^8–10^{10}) photons for every baryon in the universe? This question could presumably be answered by GUTs or SUSY by outlining the mode of creation of baryons.

There is another outstanding problem of big bang cosmology called the 'flatness' problem. This may be described as follows. The expanding space in a Friedman model can have either a uniform positive curvature, a uniform negative curvature or a zero curvature. The rate of expansion depends on this curvature property and the overall time scale associated with the universe is determined by the curvature mode in which the universe was initially set up. The mode of zero curvature (the flat mode) critically separates the other two. If the space has positive curvature it would eventually contract whereas if it has negative curvature it would quickly disperse to infinity. The adjective 'eventually' and 'quickly' are to be judged against any fundamental time scale associated with the universe. The only time scale that was available to the very early universe was the so-called Planck time

$$t_P = \frac{\sqrt{(G\hbar)}}{c^5} \approx 5 \times 10^{-44}\,\text{s}.$$

How in spite of such a short time scale did the universe manage to 'last' so

---

† It is an indication of how 'sell-oriented' science has become in today's consumer age that adjectives like 'grand' and 'super' are needed to qualify theories. Newton, Maxwell and Einstein who relied on the quality of their product than on such superlative adjectives would probably have fallen by the wayside in the race for funds.

long as its present age of $\sim 10^{10}$ years? For it to have lasted so long without contraction or dispersal to infinity much sooner, the universe has to be set up in the flat mode. How did this come about?

Then there is the 'horizon' problem arising from the fact that in a universe that has finite age, sufficiently remote parts have not had time to communicate with each other since the limit on communication speed is finite, the speed of light. At early epochs, the radius of the sphere of communication, the so-called particle horizon, is very small. For example, at $t = 10^{-37}$ s (when GUTs or SUSY operated) this radius was $\sim 3 \times 10^{-27}$ cm. Any relics of such an early era coming from regions well separated from one another should therefore show inhomogeneities. If the radiation background presently observed in the microwaves was generated very early in the universe how did it manage to acquire such homogeneity?

It is hoped that by understanding the physics of the very early universe we may be able to understand the answers to such questions. This is the motivation that drives the cosmologist to work with the particle physicist.

In a sense all the above problems may be considered relics of the very early epochs: for it is unlikely that these questions could be answered from the knowledge of the state of the universe at later epochs even if they were as early as the epochs considered by Gamow *et al.* (1–200 s) after the big bang.

There is one subtle difference between the early universe considered by Gamow and the very early universe discussed by the particle physicists and cosmologists today. To calculate the synthesis of nuclei during 1–200 s, Gamow was using laws of physics already tested in the laboratory. The reaction rates, decay rates, particle masses, the rules of statistical physics, etc. were all taken over from *known* physics. For this reason, the estimates of primordial abundances of light nuclei like helium provide a reliable check on the early universe scenario. In the case of the very early universe on the other hand, the particle physics being used has *not* been tested independently in the terrestrial laboratory; on the contrary, as emphasized earlier, it is supposed to be tested in the laboratory of the very early universe (which is itself under scrutiny!)

An analogy will help in understanding the logical status of the above procedure. Given the validity of Ohm's law, we can use it to calibrate resistances, ammeters or voltmeters. Or, given previously calibrated resistances, ammeters and voltmeters we can test Ohm's law. But we cannot do both at the same time. Likewise, if we have complete faith in the very early universe scenario we can test GUTs and SUSY; or we can take GUTs and SUSY for granted and discover what the very early universe was like. At best we may hope to arrive at a consistent picture in which a given particle theory provides a satisfactory explanation of present-day relics when used in a specific scenario of the very early universe. What we cannot claim is that our self consistent picture has *absolute* validity. It could very well be the case that *both* the particle physics and cosmology as depicted in the above solution are wrong!

Unfortunately this logical alternative is either forgotten or ignored in the definitive statements made about the origin of the universe. In the rest of this article I will discuss ways in which the above standard picture could go wrong at various levels.

The problem of the relics

An artist once exhibited a blank picture in an impressive frame. When asked what the picture was, he said that it was about a cow grazing. To the question as to where was the grass his answer was that the cow had finished it all. Where was the cow? The cow had left because it could no longer find grass there!

The scientific analogue of the above story is the following. Make a hypothesis H_1 to predict the existence of a certain relic R. If you don't find R, make another hypothesis H_2 to argue that after creation R could not have survived long enough to be seen. Thus the non-observation of R allows you (apparently!) to confirm the validity of two hypotheses H_1 and H_2.

We see examples of this type in the so-called relics of the very early universe. Some GUTs models predicted the existence of massive magnetic monopoles which are not only *not* seen today but whose existence (if detected) would be positively embarrassing for cosmology. For example, a typical monopole would have a mass of $\sim 10^{-8}$ g and the expected number density of such monopoles would be $\sim 10^{-6} \, \text{cm}^{-3}$. Thus the mass density of monopoles of $\sim 10^{-14} \, \text{g cm}^{-3}$ would be far higher than the *total* mass density of $\sim 10^{-29} \, \text{g cm}^{-3}$ predicted for the universe by most big bang models! Moreover the existence of free monopoles would have played havoc with the magnetic field in our Galaxy and outside it. So as to get rid of the unwanted monopoles a new hypothesis is needed.

The existence of massive neutrinos is another example. GUTs and SUSY predict a variety of new particles of which neutrinos are expected to survive from the very early epochs, because they only weakly interact. However, massive neutrinos tend to lower the age of the universe (a problem I will come to later), introduce too much patchiness in the distribution of galaxies and possibly destroy the good agreement between the observed and calculated values of primordial helium. So numerous constraints are needed to ensure that most of the massive neutrinos do not survive for long.

The problem of monopoles and neutrinos has acquired additional interest because there are indications that the universe might possess non-luminous ('hidden') matter in a substantially greater measure than the luminous part in the form of galaxies. Could this matter be in the form of black holes or made up of exotic particles like photinos, gravitinos, axions etc?

It is certainly not beyond the combined ingenuity of particle physicists and cosmologists to produce scenarios in which the awkward monopoles do *not* get created, the massive neutrinos do *not* survive and the more exotic particles fulfill the necesssary requirements of hidden matter by remaining unseen. The overall self-consistency of this picture, as and when it eventually emerges, can at best be *a* plausible demonstration of how the universe began: it can never be a scientific proof that this was *the* way the universe began.

The best bet for a consistent scenario of the very early universe is offered at present by the so-called inflationary model. Here too, the original idea proposed in 1981 by A. Guth, though elegant and ingenious, did not work. It had to be replaced by the 'new' inflationary model which had its own crop of difficulties. One of the difficulties currently faced by the inflationary models is the predicted very large density fluctuation in the distribution of matter and radiation as seen today in the universe. The observed small scale smoothness of the microwave background clearly contradicts this prediction. Here too, the difficulty may be eliminated in an inflationary model, Mark n ($n > 2$), by exploiting some hitherto unexplored parameters, of a particularly exotic particle theory.

I think the present day cosmologists should pause now and then to ask themselves if they are going the same way that their ancient Greek counterparts did two millennia ago. Are the parameter fitting exercises needed to prop up a given scenario for the very early universe any different from the epicycles of Hipparchus and Ptolemy?

I end this section by discussing a difficulty of the inflationary model that will not go away by any amount of juggling with the parameters of GUTs and SUSY. The model, while 'resolving' the flatness problem also predicts that the age of the universe must be given by $\frac{2}{3}H_0$, where H_0 is the present value of the Hubble constant. The Hubble constant is measured observationally, as the ratio of the redshift (z) of a relatatively nearby galaxy multiplied by the speed of light (c) to its distance (D):

$$H_0 = \frac{cz}{D}.$$

The values of H_0^{-1} are considerably uncertain, lying in the estimated range of $10^{10} - 2 \times 10^{10}\,\mathrm{yr}$. There are observational reasons in favour of both values but the 'middle of the way' consensus among the astronomers today is to take the value at $\sim \frac{4}{3} \times 10^{10}\,\mathrm{yr}$.

Corresponding to this value the age of the universe is $\sim 9 \times 10^9\,\mathrm{yr}$. This value is too small to accommodate the age of the Galaxy and in particular the ages of globular clusters. These ages are variously estimated at $(10–18) \times 10^9$ yr. Even taking Hubble's constant at the lowest value in the above range, the predicted age of the universe is not able to exceed the estimated ages of various astronomical systems in it.

This difficulty is usually pushed under the rug by taking $H_0^{-1} = 2 \times 10^{10}$ yr and stating that this is the 'Hubble age' of the universe. Nothing can be more misleading than this statement. If one is to take the current band-wagon of the very early universe seriously, then there is no escape from the conclusion that the 'true age' of the universe is two thirds of the 'Hubble age' and that it falls woefully short of other astronomical ages.

Quantum cosmology

Maxwell's electromagnetic theory coupled with the Lorentz–Einstein inputs of special relativistic electrodynamics admirably describes macroscopic phenomena. Yet the theory is found to be inadequate for microscopic purposes like, for example, the study of the simplest atom, that of hydrogen.

Classical electrodynamics tells us that the hydrogen atom should have a very short term existence. For, an electron circling a proton radiates continually and this energy loss brings it closer and closer to the proton until it merges with it. The time scale for this phenomenon is none other than the classical electromagnetic time scale $e^2/mc^3 \sim 10^{-23}$ s where e and m are the charge and mass of the electron. That the hydrogen atom is of a stable nature and that whenever it radiates it does so in discrete packets of energy rather than in a continuous manner are sufficient to tell us that the classical description is inadequate.

Quantum theory solves this problem in a satisfactory way. Given the Planck's constant, the mass of the electron and the electronic charge, we can construct a length scale $\hbar^2/me^2 \sim 10^{-8}$ cm, at which quantum ideas become relevant. Detailed atomic theory then tells us that this is the characteristic size of the atom and that the discreteness is the result of stationary states.

The existence of stationary states could be discovered even without going into the full details of quantum theory by simply concentrating on the quantization of the radial separation r between the electron and the proton. Such an approach does not tell us about the numerous stationary states due to the angular momentum quantum numbers. But it gives us the crucial information that the 'singular' fate of the classical H-atom arising from $r \to 0$ is avoided by its quantum counterpart.

All this does have relevance to cosmology! Like the Maxwell–Lorentz–Einstein electrodynamics the general theory of relativity is also a successful theory of gravity at the classical level. It has, of course, not been tested in the situations where gravity is very strong and concepts such as the 'black hole' and the 'big bang' are logical extrapolations of the theory into untested domains.

While physics in general would not progress but for such extrapolations, the example of the classical H-atom warns us to be cautious. In particular, can we trust the conclusion of classical general relativity that the big bang origin of the universe is inevitable? Can we trust the theory at time scales less than the Planck time t_P mentioned earlier? Was there a big bang at all, according to quantum theory?

Obviously the quantum theory of gravity holds the key to this important mystery. Unfortunately, in spite of many continuing attempts, a formally satisfactory and at the same time practically workable theory of quantum gravity still remains unattainable. Can we, however, expect to capture the flavour of the quantum inputs by a less ambitious but mathematically more manageable approach?

Such an approach is fortunately available. Just as the essence of the problem (but not the full details) could be captured by quantizing r for the H-atom, we can similarly try to quantize the scale factor S of the expanding universe: for, the classical big bang singularity is obtained by letting S go to zero.

While S can be quantized, a more convenient way (and also physically a more satisfactory one) is offered by the method of *conformal quantization*. According to this picture, from any classical spacetime geometry that satisfies Einstein's equations we can generate new geometries by arbitrarily scaling all spacetime intervals at every point by a factor Ω. This factor, known as the conformal factor, could change from one spacetime point to another. By quantizing Ω we therefore obtain a fairly general description of how scale fluctuations of spacetime occur around the Planck epoch.

Recent work by myself and T. Padmanabhan shows that such quantum fluctuations describe a host of spacetime geometries, the vast majority of which are *not* singular. In other words, the classical big bang turns out to be more the exception than the rule that characterizes the pre-Planck-time universe. The quantum mechanical probability that the universe arose from a big bang is almost zero. Thus the notion of an 'origin' a finite time ago is almost ruled out. Instead we have a universe that exists for ever and whose behaviour is determined largely by classical gravity. The exceptions arise when it happens to pass through highly compact states when quantum gravity comes into play. In the quantum regime there will be transitions of the universe from one state to another until it again expands and emerges in a classical state.

The removal of the concept of 'origin' also removes the horizon problem since there is now no limit on the range of communication. Likewise, if we compute the probability that through quantum conformal fluctuations the universe got into a Friedman–like model from an initial empty (vacuum) state then we find that the flat Friedman mode is overwhelmingly preferred. Thus the universe is without a singular origin and without the problems of horizon and flatness that beset the classical big bang models.

So it appears that there is a prima-facie case for quantum cosmology significantly altering the conclusions of classical cosmology. If the problems of singularity, horizon and flatness are resolved in the quantum era, then much of the motivation for the inflationary phase occurring later disappears.

Is the microwave background a relic of the big bang?

I next turn to another non-standard concept that questions the main evidence for the hot big bang. The concept has roots in the question posed above.

The main expected signatures of the microwave background as relic radiation were the following:

(i)   Its spectrum should be Planckian.
(ii) It should show small scale fluctuations that relate to the era of galaxy formation.
(iii) Its energy density should be deduced from the physical conditions prevailing soon after the big bang.

Had these signatures been confirmed there would have been no difficulty in accepting the microwave background as the relic of the big bang. In reality the situation has been otherwise: only the first signature appears to be confirmed.

The extraordinary smoothness of the radiation background poses problems for scenarios of galaxy formation. If galaxies formed after the radiation background was made, why did the process of their formation leave no apparent marks on the background? Here again, after the realization that the original and more natural theory of galaxy formation through an adiabatic process leaves too large a fluctuation on the background, the less persuasive isothermal process was invoked. The latter leaves so small a fluctuation of temperature that it cannot be detected in the foreseeable future. This certainly saves the scenario, but at the same time it deprives it of predictive vulnerability that is the hallmark of a good scientific theory (this is another example of the cow grazing scenario!)

The energy density of the radiation, at the present temperature of 3 K is $\sim 6 \times 10^{-13}\,\mathrm{erg\,cm^{-3}}$. Why is the present temperature ~ 3 K? Why not 1 K or 10 K? Clearly, according to the relic radiation hypothesis the answer to this question must lie in the history of the very early universe. So far this answer has not come from the current ideas in big bang cosmology.

This is where one is tempted to look elsewhere for the origin of this background. Could it have been of a much recent origin, having nothing to do with a hot big bang? There are several astrophysical processes (of no connection with cosmology) that do produce energy densities of this order. To name a few, the galactic magnetic field, the cosmic rays and starlight all produce energy densities of the same order. With regard to the starlight, it was pointed out in 1968 by F. Hoyle, N.C. Wickramasinghe and V.C. Reddish that if all helium observed in the universe were made in stars, the starlight so generated would have the same energy density as the microwave background.

This last coincidence suggests a stellar origin of this background provided it can be thermalized subsequently. M.J. Rees proposed in 1978 that Population III stars at the epoch of $\sim 10^7 - 10^8$ yr could process the helium. Even earlier, in 1974–75 Wickramasinghe and others (including this author) had proposed that even ordinary starlight in more recent epochs would be efficiently thermalized by long slender grains of graphite. The details of such a process were studied extensively by N.C. Rana in 1980–81, who found that a plausible model of the microwave background can be made up within the observational constraints.

It is early days yet to assess such attempts, but their value as alternatives to the hot big bang scenario (since it is no longer as attractive as before) cannot be discounted.

How universal is Hubble's law?

I have discussed alternatives to the standard hot big bang cosmology, in an order that is progressively more radical. The last in this series questions the very basis of the idea of the expanding universe. As is well known, the redshift/distance relation of Hubble applies to galaxies over a wide range of apparent brightness. But, the notion of the expanding universe implies that Hubble's law must apply to *all* extragalactic objects.

We may express the above requirement thus. Given an extragalactic object at distance D, its redshift should ideally be given by a *unique* relation of the kind

$$z = f\left(\frac{DH_0}{c}\right)$$

where f is a function determined by the cosmological model. For small distances $f(x) \sim x$ and the above formula reduces to the linear relation first found by Hubble. However, *whatever* be the cosmological model, all objects at the same distance must have the same redshifts.

In practice small departures from this idealized situation are permitted, largely because galaxies in a cluster may have random motions of the order of $\leqslant 1000 \, \mathrm{km \, s^{-1}}$. Translated into spectral shifts by the Doppler formula, these motions may generate departures in z as given by the Hubble law, of the order of $\Delta z \sim 0.003$. Indeed historically speaking, the very first example of extragalactic spectral shift to be investigated was one of blueshift!

Barring such small fluctuations therefore we expect all extragalactic objects to obey the above generalized Hubble relation. Do they?

Over the last two decades there has been a steady accumulation of data that seem to cast doubts on the universality of the Hubble relation. The data have been collected by many observers but by far the lion's share comes from the observations of H.C. Arp. The data are of the following kind.

    (i)   There are pairs or larger groups of quasars that appear to be near neighbours in space but with individual members having very different redshifts.

   (ii)   There are quasar-galaxy associations wherein typically a high redshift quasar is found near a low redshift galaxy.

  (iii)   There are galaxy–galaxy associations wherein typically a companion galaxy with larger redshift (vastly exceeding $\Delta z \simeq 0.003$) appears to be dominated by a main large galaxy of lower redshift.

It should be mentioned that all these cases are at present highly controversial, the dispute arising because we do not know the distances of the objects. There is no independent way of measuring distances, especially of the quasars concerned. Physical nearness of two quasars or of a quasar and a galaxy is usually argued on the basis of their projected nearness on the sky. Statistical arguments are needed to decide whether two objects at vastly different distances from us would happen by chance to be projected very near each other as seen by us. If the probability of chance projection is very low (say $< 10^{-2}$) then there is reason to suspect that the objects in question are physically near each other. The controversy is centred round the way the probabilities are computed.

These statistical arguments apart, Arp has produced in case (iii) examples of filamentary structures linking galaxies with discrepant redshifts. Are these structures proofs of physical connection between the galaxies? It seems hard to discount such evidence as 'projection effect' or 'photographic artefacts'.

Quasars being highly unusual objects might have other causes for their redshifts; but to argue that Hubble's law does not hold for galaxies would shake the very foundations of modern cosmology.

Conclusions

Nonstandard cosmology, by definition, includes ideas not conforming to the standard hot big bang model and as such it includes much more than what is presented here. For example, it is possible to work in the frame of other gravity theories, to assume that fundamental constants are changing or to propose completely new laws of physics.

To avoid entering into an open-ended field I have concentrated on motivations provided by observational cosmology. The aim here was not to offer a cut and dried alternative to the standard hot big bang but to emphasize that the standard model by no means provides the last word on the origin of our universe.

9

The dark matter problem

BERNARD CARR

1 Introduction

One of the most remarkable discoveries of modern astronomy is the fact that only a small fraction of the Universe is in visible form. At least 90% of its mass is dark and a prime challenge in cosmology today is to determine the nature of this dark component. The first observational hints of the existence of dark matter came some 50 years ago but only in the last decade have astronomers come to realize how pervasive it is and only recently have theoreticians begun to speculate about all the possible explanations for it.

In a sense the pervasiveness of dark matter should occasion no surprise. We know that visible light only corresponds to a small part of a spectrum of electromagnetic radiation, as illustrated in Fig. 9.1. There is nothing particularly special about it except that it corresponds to the waveband in which our eyes happen to be receptive. The fact that astronomers were initially surprised by the dark matter problem doubtless reflects Man's continual tendency towards anthropomorphism. Actually, most things in the Universe are dark (at least in the visible range) and objects which radiate (such as stars and hot gas) only do so for a limited period. One should not be surprised therefore if dark matter takes on as many different forms as visible matter. Indeed, the challenge posed by the dark matter problem is not so much to generate solutions as to eliminate them.

What makes the identification of the dark matter a particularly exciting task is the fact that it involves so many different fields of physics. On the theoretical front, it involves general relativity, high energy physics, unified field theories, and almost every branch of astrophysics; on the observational front, it involves the search for gravitational waves, particle physics experiments, and observations over almost every waveband from gamma-rays to radio waves. It thus requires the sort of corporate scientific enterprise which

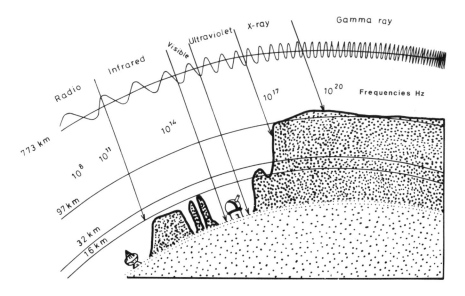

Fig. 9.1. Transparency of atmosphere to radiation. This shows the penetration depth into the Earth's atmosphere of the various kinds of
electromagnetic radiation. Only visible light and some of the radio
band reach the Earth's surface. The other wavebands are unable to
do so because of absorption in the upper atmosphere by ozone,
oxygen and water vapour.

history shows has always led to important advances and cross-fertilization of
ideas.

The dark matter problem is not only of interest because of its cosmological
significance. It is also intriguing because of the way in which cosmologists
approach it. The history of the problem seems to reflect many of the features
of scientific progress in general. For several decades the scientific community
was reluctant to take the problem seriously; now, in a comparatively short
time, it has become very excited about it. The excitement comes from two
directions. On the one hand, it comes from the particle physicists – already
attracted to cosmology because of the interface the physics of the early
Universe affords with their own subject. This group of people, naturally
enough, are enthusiastic to explain the dark matter problem in terms of
elementary particles. On the other hand, it comes from the astrophysicists –
already long absorbed in cosmological speculation – who prefer to attribute
it to something astrophysical. By and large, the particle physicists are more
influential because there are more of them and they can attract more funding.
But the astrophysicists can also boast some influential champions.

Under the influence of these two groups, theories for the dark matter have multiplied with alarming rapidity. Many of them only last a few years (except perhaps in the minds of their original advocates) but, in general, theories are created faster than they are destroyed. Occasionally one particular theory may become a front runner as a result of receiving temporary stimulus from some observational result, only to be relegated to the ranks of the other runners when the result is not confirmed by subsequent observations. As the bandwagon of cosmological speculation veers from side to side under the competing driving force of the particle physicists and astrophysicists, the more sedate scientist might be forgiven for wondering whether it is going in the right direction at all.

Such pessimism is almost certainly unwarranted. The particle physics/astrophysics interface has seen tremendous progress in the last decade (through both observational and theoretical ingenuity) and it would be perverse to assume that this progress will not continue. We can probably have confidence that the bandwagon is at least going roughly in the right direction, even though we will have to wait for the dust to settle before we can see where that direction leads. My own hunch is that the resolution of the dark matter problem (which, as we will see, really breaks down into several distinct problems) will require both astrophysics and particle physics. At this stage it seems rather unlikely that the solution will involve any fundamentally new physics (although this has been proposed). More likely, it will involve some of the revolutionary ideas which the last decade of physics has already brought (for example, the existence and evaporation of black holes, the non-conservation of baryon number, the possibility of magnetic monopoles, the non-zero rest mass of the neutrino). Nevertheless, solving the dark matter problem is bound to expand Man's perspective of the world.

Scientific progress, of course, always does involve an expansion of perspective and it is perhaps worth contemplating some of the levels on which this expansion proceeds. One such level involves the question of scale. While our everyday experience is confined to scales between millimetres and kilometres, the development of progressively sophisticated instrumentation has extended the range of scales we can experience enormously: particle accelerators now allow us to probe structure on scales down to 10^{-15} cm, while modern telescopes allow us to peer to the edge of the visible Universe (10^{28} cm). Thus technological progress has extended our domain of experience from 6 decades of scale to 43. Intellectual developments have extended the range still further: theories of quantum gravity speculate about structure on the scale of the Planck length (10^{-33} cm), while the latest cosmological theories regard the visible Universe as just a tiny bubble in a still vaster cosmos. The solution to the dark matter problem could be relevant to processes involving both of these extremes.

Another level on which our widening of perspective proceeds concerns the way in which we observe the Universe. In everyday experience most of our information of the world comes via light but, as mentioned earlier and indicated in Fig. 9.1, there are many other wavebands in which we can procure information. Atmospheric effects make it impossible to study the Universe in most of these wavebands from the ground but the advent of space telescopes now allows us to remedy this. Nor is our perception of the Universe any longer confined to electromagnetic waves. The advent of gravitational wave detectors and neutrino detectors should soon open yet further windows on the Universe, permitting us to witness processes which would be quite undetectable with any form of electromagnetic radiation. All of these techniques may be relevant to solving the dark matter problem.

After these general remarks, I will now turn to a detailed discussion of the dark matter problem itself. Firstly, in Section 2, I will review the evidence for the dark matter, emphasizing the several distinct contexts in which it arises. Then, in Section 3, I will discuss the many explanations which have been advanced for it, some proposed by astrophysicists and some by particle physicists. In Section 4, I will consider the various constraints on the form of the dark matter, illustrating why these severely restrict the number of viable theories. In concluding, in Section 5, I will emphasize that one should not expect any single explanation to solve all the dark matter problems and that what is required is probably a synthesis of astrophysical and particle physical explanations.

2   Evidence for dark matter

On astronomical scales any stable structure in the Universe reflects a balance between the attractive effect of gravity (which tries to make it collapse) and the repulsive effect of some form of the pressure (which tries to make it expand). This balance is expressed by what is called the virial theorem: this states that for a bound system of mass M and radius R, there is a characteristic velocity $V \sim \sqrt{(GM/R)}$, where G is the gravitational constant. (This order-of-magnitude relation can be specified more precisely in any particular context, but the details need not concern us here.) For a system of stars or galaxies, this velocity reflects either a random motion (like the 'thermal' motion of the atoms in a gas) or a systematic motion (like the rotation of the planets around the Sun). In either case, the quantities V and R can be measured, the first via the Doppler effect and the second by means of some sort of distance indicator, thereby leading to an estimate of M.

A dark matter problem arises whenever the amount of gravitating mass implied by the virial theorem appears to exceed the amount of mass in visible form. The associated dark matter is sometimes referred to as 'missing mass', although strictly it is the light and not the mass which is missing. There is evidence for dark matter in at least four different contexts. We will discuss them in order of increasing scale, although this is not the order in which they were discovered historically.

The first kind of dark matter is associated with *the disk of our own galaxy*. In this case, the relevant value of V is the speed with which the stars in the disk move up and down perpendicular to the disk (observed to be about 10 km/s) and the relevant value of R is the disk thickness (about 100 pc). It has been known since the 1940s that the disk density determined in this way exceeds the density observed in gas and visible stars. The most recent calculations indicate that 50% of the disk mass is dark. Although this is a fairly modest fraction (compared to that associated with the other dark matter problems), it is the dark component for which the evidence is most unambiguous. The observations also indicate that the disk dark matter must itself be confined to the disk; it does not have a spheroidal distribution.

The second kind of dark matter is associated with *the halos of spiral galaxies*. In this case, the relevant value of V is the speed with which the stars rotate around the centre of the galaxy and the relevant value of R is their distance from the centre. The way in which V varies with R specifies what is called the 'rotation curve' of the galaxy. Once this curve is measured, the virial theorem allows one to infer how the mass within R (and hence the average density within R) varies as one moves outwards. In several dozen spiral galaxies, the rotation speed appears to be constant (usually of order 100 km/s) as far as the visible stars extend. This corresponds to a density which falls as R^{-2} (or to a mass which increases as R), whereas the density of visible stars falls off much faster than this.

At the edge of the visible galaxy (typically $R \sim 10$ kpc), the value of M is not much larger than the visible mass (usually of order 10^{11} M$_\odot$). However, in many cases the rotation curve can be measured out to distances well beyond the visible stars (e.g. by making 21 cm observations of neutral hydrogen) and it still remains constant. This indicates that spirals have a dark component which extends much further than the visible material and contains considerably more mass. For our own galaxy, the kinematical properties of globular clusters and neighbouring galaxies suggest that the dark material extends to at least 30 kpc. Independent evidence for the existence of halos comes from the form of the disks in spiral galaxies: in particular, from the persistence of warps and from the fact that an extended halo may be required to stabilize disks against the formation of 'bar' structures. Both these features require that the halo dark matter have a spheroidal distribution, so it must be different from the dark matter in the disks themselves. The total mass associated with a typical halo depends on the radius R_H to which it extends (which is unknown). If $R_\mathrm{H} = 50$ kpc, corresponding to a radius typically five times bigger than that of the visible galaxy, it would be about $M \sim 10^{12}\,\mathrm{M_\odot}$.

The third type of dark matter is associated with *clusters of galaxies*. This is actually the oldest dark matter problem, having been discovered in the 1930s. In this case, the relevant value of V is the random speed of the galaxies within the cluster (usually about $10^3\,\mathrm{km/s}$) and the relevant value of R is the radius of the cluster (about 10 Mpc). Measurements in rich clusters indicate that their total mass ($M \sim 10^{15}\,\mathrm{M_\odot}$) exceeds the mass in the visible galaxies by at least a factor of 10. The dark matter cannot be gas since it would have to be very hot ($10^8\,\mathrm{K}$) to avoid sinking into the middle; it would then produce far more X-ray emission than is observed. (However, some X-rays are seen and this suggests that the gas density is at least comparable to that in galaxies.) The dark matter in clusters could in principle be the same as that in galactic halos. Indeed, in the 'hierarchical clustering' picture (in which, as time proceeds, ever larger scales of cosmic structure form as a result of gravitational clumping), one would expect the galaxies inside a cluster to be at least partially stripped of their individual halos, thus forming a collective dark component. However, this would explain the amount of dark matter in clusters only if the original galactic halos had a sufficiently large value of R_H, which is uncertain.

The fourth kind of dark matter is associated with a *smooth cosmological background*. A crucial issue in the Big Bang theory is whether the Universe has more or less than the critical density required for its eventual recollapse. The visible material in galaxies has only about 1% of the critical density and even the dark material in galactic halos or clusters could have only 10–20%. However, there could still be another dark component which makes up the remaining 80–90%, providing it is less clustered than the galaxies themselves.

There is no observational evidence for this component and its existence used to be postulated for purely aesthetic reasons (e.g. to satisfy Mach's principle). Recently, however, a more compelling theoretical reason for anticipating a critical density has been advanced. The idea is that the early Universe may have undergone a period of extremely rapid expansion (called 'inflation') as a result of a cosmological phase transition. (In the usual Big Bang theory the scale factor increases only as a power of time, whereas it increases exponentially during inflation.) This theory has a number of attractive features: in particular, it explains why the Universe looks so smooth, why there are not an excessive number of magnetic monopoles (as discussed later), and why the initial density fluctuations arise. A crucial prediction of the theory is that the total density must have almost exactly the critical value. One would naively presume that the extra dark matter cannot be the same as the dark matter in clusters (since the latter does cluster by definition). However, one could circumvent this conclusion by invoking what is termed a 'biased galaxy formation' scenario, in which galaxies form preferentially in just a small fraction of the volume of the Universe.

We may summarize this section by saying that at least 90% and, if one accepts inflation, 99% of the Universe's mass is in some dark form. We have identified four contexts in which dark matter seems to be necessary. It is interesting that the dark matter density required seems to increase systematically as one goes to larger scales. In principle, all four sorts of dark matter could be different, even though an Occam's razor argument might suggest otherwise. For example, we have seen that the cluster dark matter could be the same as the halo dark matter only if the initial halo radius was large enough; and the background dark matter could be the same as the cluster dark matter only if one invokes biased galaxy formation. Therefore, in discussing the various explanations for the dark matter (as we do in the next section), one must always be careful to specify exactly which dark matter problem is being considered.

3 Candidates for the dark matter

There are many possible explanations for the various forms of dark matter discussed above. As mentioned in the Introduction, this is perhaps not surprising when one realizes that most things in the Universe are dark. A crucial question is whether the dark material consists of the sort of ordinary (baryonic) matter which makes up visible objects or of something more exotic. The exotic (non-baryonic) candidates correspond to some type of elementary particle (generically called an 'ino'); such particles are supposed to be relics of processes which occurred in the early Universe (usually the first second). The baryonic candidates are supposed to form out of the background gas at a relatively late stage (viz. 10^7-10^9 yr after the Big Bang) through astrophysical mechanisms; this may be termed the 'Population III' scenario since it involves the formation of a lot of stars, analogous to the 'Population I' and 'Population II' stars seen in galaxies but forming at an earlier epoch. Another candidate, the primordial black hole, is supposed to form from gravitational collapse in the early Universe. This candidate will be included in the non-baryonic category (even though the black holes concerned may be large enough to be regarded as astrophysical) for reasons which will become clear later. These possibilities are illustrated qualitatively by the first two diagrams in Fig. 9.2 and the candidates are listed explicitly in Table 9.1.

We first consider the *elementary particle candidates*. In the conventional hot Big Bang picture, the temperature of the background radiation increases continuously as one goes back in time. This means that any type of particle will be roughly as numerous as the background photons once the temperature exceeds the particle's rest mass. Furthermore, since all interaction

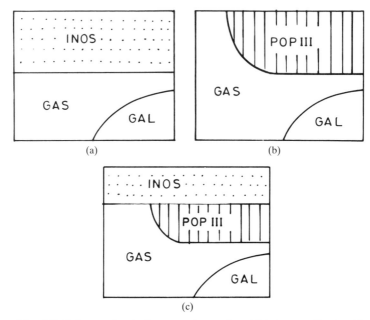

Fig. 9.2. Schemes for dark matter formation. This is a qualitative illustration of three scenarios for dark matter formation. In (*a*) the dark matter is contained in 'inos' which are a relic of the big bang; in (*b*) the dark matter is contained in 'Population III' objects which formed out of the background gas; (*c*) is a compromise scenario in which the background dark matter consists of inos but the halo and perhaps cluster dark matter consists of Population III objects. In all three scenarios, galaxies form out of the remaining background gas after the dark matter has formed.

Table 9.1. *Dark matter candidates. This lists the various elementary particle and astrophysical candidates which have been proposed for the dark matter. The candidates are listed in order of increasing mass, the mass (where known) being indicated in parentheses. All these masses are very approximate:* $1\,M_\odot \approx 10^{33}\,g$, $1\,eV \approx 10^{-33}\,g$

Inos		Population III	
Axions	$(10^{-5}\,eV)$	Snowballs	?
Neutrinos	$(10\,eV)$	Jupiters	$(<0.08\,M_\odot)$
Gravitinos	$(1\,keV)$	M-dwarfs	$(0.1\,M_\odot)$
Photinos	$(1\,GeV)$	White dwarfs	$(1\,M_\odot)$
Monopoles	$(10^{16}\,GeV)$	Neutron stars	$(2\,M_\odot)$
Shadow matter	?	Stellar black holes	$(10-10^5\,M_\odot)$
Primordial black holes	$(>10^{15}\,g)$	Supermassive black holes	$(>10^5\,M_\odot)$

rates involving the particle increase rapidly with the temperature, these rates will exceed the cosmological expansion rate before some critical time t_F. Prior to t_F the particle should therefore exist in thermal equilibrium with the background photons but after t_F it should 'freeze out'. Provided it is relativistic at this time (i.e. provided its rest mass m_x is less than the freeze-out temperature T_F) and provided it survives until the present epoch (i.e. provided it does not decay), its present number density should just be $n_x \sim g n_\gamma \sim 100 g$ cm^{-3}, where n_γ is the number density of the microwave background photons and the factor g arises because the annihilation of other particle species after the freeze-out time will increase the relative photon density. (Thus, as T_F increases, g decreases in a series of steps, each step corresponding to the rest mass of some particle.) If the particles are non-relativistic today, their present mass density is just proportional to m_x; in units of the critical density it should be $\sim g(m_x/100\,\text{eV})$.

The original elementary particle candidate was the *neutrino*. This freezes out when the weak interaction rate for processes involving neutrinos falls below the expansion rate at $T_F \sim 1\,\text{MeV}$ (corresponding to $t_F \sim 1\,\text{s}$); in this case $g = \frac{3}{11}$. Thus the neutrino could be of cosmological significance if its rest mass m_ν exceeds about 10 eV. (It used to be assumed that m_ν is zero, but a value of order 10 eV is not inconsistent with 'unification' theories.) Attention was focussed on this possibility as a result of Russian studies of tritium decay, which appeared to indicate a neutrino rest mass in the range $14\,\text{eV} < m_\nu < 46\,\text{eV}$. However, this is a very difficult experiment and the results have yet to be confirmed. Independent evidence could come through the detection of neutrino oscillations: electron neutrinos of energy E should turn into other types of neutrinos over a distance $\sim 10(E/10\,\text{MeV})\,(m_\nu/10\,\text{eV})$ metres. At one stage there were claims to have found such an effect over a distance of 10 metres for neutrinos from nuclear reactors ($E \sim 10\,\text{MeV}$). However, later experiments have not confirmed this. Indirect evidence for neutrino oscillations may come from attempts to detect neutrinos from the Sun (which also have $E \sim 10\,\text{MeV}$): the observed flux of electron neutrinos appears to be about a third that expected from nuclear reactions within the solar core and this may be explained rather naturally if there are three neutrino species which continually oscillate into each other (thereby reducing the flux of any particular species by a factor of three). However, the oscillation lengthscale is less than the distance to the Sun (as required) for any value of m_ν exceeding $10^{-6}\,\text{eV}$, so the solar neutrino problem itself does not require that m_ν be large enough to be of cosmological significance.

Another possibility is to invoke a particle whose mass is larger than 100 eV but which decouples earlier (reducing g) so that its density is not too large. Candidates for such a particle arise in the supersymmetry theories (for which $T_F \sim 10^{10}\,\text{GeV}$ and $g \sim 10^{-2}$) and include the *gravitino* (which has a mass of about 1 keV). Some of the supersymmetry particles have such a large mass that they will be non-relativistic when they freeze-out. In this case, many of the particles are expected to annihilate with each other, so n_x is no longer related to n_y in the simple way indicated above. Nevertheless, it turns out to be rather natural for these more massive particles to have around the critical density. This applies, in particular, to the *photino*, which has a mass of order 1 GeV.

Although there are a large number of different 'ino' candidates, they can be conveniently classified into three groups. This is because the present momentum of any relic particle should just be of order the momentum of a microwave background photon. This implies that its speed – prior to any clustering – should be inversely proportional to its mass. The slowest moving most massive particles (like the photino) are therefore termed 'cold', while the intermediate mass particles (like the gravitino) are termed 'warm', and the lightest particles (like the neutrino) are termed 'hot'. The distinction is important because we will find that the scale on which a particle can cluster depends upon how cool it is.

Another 'cold' candidate which is currently enjoying popularity is the *axion*. This particle is associated with an extra symmetry introduced so that strong interactions do not exhibit CP violation (i.e. the interactions must look the same under simultaneous reversal of charge and parity). This symmetry is spontaneously broken at an energy scale f_A which astrophysical constraints require to be of order 10^{12} GeV. At a lower temperature of about 1 GeV, the axion develops a mass m_A due to 'instanton' effects. The associated density parameter is $\sim (m_A/10^{-5} \text{eV})$ in units of the critical density. Although the value of m_A is not known precisely, it could certainly be of order the value 10^{-5} eV required for the density to be significant. Note that the axion is the lightest of the ino candidates; the fact that it is colder than the neutrino, even though it is lighter, reflects the different way in which it arises.

The most massive elementary particle candidate is currently the *magnetic monopole*. It used to be thought that the existence of such a particle is forbidden. However, it is a remarkable prediction of the 'grand unified theories' that monopoles can exist. Indeed, in many models they have to form at the 'grand unification' epoch (around 10^{-35} s after the Big Bang) as a result of a phase transition. The problem is that they are so massive ($m_M \sim 10^{16}$ GeV) and so numerous (roughly one per horizon volume at their formation epoch) that a naive calculation suggests their density today would be impossibly large. The only way around this problem seems to be to invoke the inflationary scenario (discussed in Section 2) since this dilutes the monopole number density. But in this case, unless the amount of inflation were very finely tuned, their present density would be negligible. Furthermore, if monopoles did have a critical density, they would be expected to have destroyed the galactic magnetic field. One experimenter does actually claim to have detected a monopole but this claim is not supported by other experiments.

Doubtless theorists will concoct many more types of elementary particles in the years to come, many of which might in principle have survived as relics of the Big Bang, so it is too difficult at this stage to assess the front runner. The latest candidate, for example, is *shadow matter*. This is matter which resembles ordinary matter but only interacts with it gravitationally; its existence may be predicted by 'superstring' theory (the latest unification model). It is probably premature to lay bets on any particular candidate. However, with such a large zoo of inos, it is not implausible that at least one of them will turn out to be cosmologically significant. A particularly exciting prospect is that we may soon be able to search for some of the ino candidates using terrestrial detectors. This should at least enable us to eliminate some of the possibilities.

The only remaining non-baryonic candidate is the *primordial black hole*. Such holes could have formed from inhomogeneities or phase transitions in the early Universe. At their formation epoch, they would need to have of order the mass within the particle horizon, which is $10^5 (t/s) M_\odot$, so they could span an enormous range of masses. Those forming at the Planck time ($\sim 10^{-43}$ s) would have a mass of only 10^{-5} g, whereas those forming at 1 s would be as large as $10^5 M_\odot$. Primordial black holes smaller than 10^{15} g, which form before 10^{-23} s, are of great theoretical interest because they are the only ones small enough for quantum effects to be important. Indeed such holes would have completely evaporated by the present epoch, most of their mass having been channelled into gamma-rays, so only ones larger than 10^{15} g could solve any of the dark matter problems. In fact, the observed density of the gamma-ray background implies that 10^{15} g black holes could only have had a tiny fraction of the critical density (at most 10^{-8}). This suggests that primordial black holes can solve the dark matter problem only if they form at a phase transition after 10^{-23} s. One possibility is that primordial black holes could have formed prolifically at the quark–hadron phase transition, 10^{-6} s after the Big Bang. In this case, they would have the mass of Jupiter and a size comparable to a football. (In a way, this is the most attractive dark matter candidate since it is the only one we could ever hold in our hand!)

We now consider *Population III* candidates for the dark matter. The idea here is that the form of the initial density fluctuations may have allowed a large fraction of the Universe to have gone into bound clouds before galaxies themselves formed. For example, in the 'hierarchical clustering' scenario, the first bound objects would have a mass of about $10^6 M_\odot$ and would bind some time in the period 10^6–10^8 yr. In the 'pancake' scenario, in which the first objects to form are of supercluster scale, the pancakes could fragment into $10^6 M_\odot$ clouds before these clouds themselves reassemble to make galaxies; in this case the clouds would be forming at about 10^9 yr. In either case, one would expect the clouds to collapse and fragment into smaller Population III objects. Provided these objects are in a mass range such that they can become dark, one could in principle generate a lot of dark matter prior to galaxy formation.

This scenario requires that the first bound clouds turn into Population III objects of the right sort with high efficiency. At first sight, both these requirements might seem rather implausible. However, we do know of situations in the present epoch where gas turns into stars with high efficiency (e.g. in starburst galaxies or cooling flows). The more tricky issue is whether one could expect the Population III objects to have a mass such that they become dark. This is very hard to predict *a priori* since it depends on the mass-scale at which fragmentation within the original cloud ceases. The suggestions range from objects as small as snowballs to objects as large as supermassive stars. Admittedly, the stars forming at the present epoch do not span this range of masses, but it is not implausible that the first stars would have been rather different. We will discuss each of the possibilities in turn.

Snowballs of condensed hydrogen have been proposed, although these are probably the least likely candidate. In order to have avoided collisions within the age of the Universe, they must have a size of at least 1 cm but they would then have been evaporated by the microwave background radiation. In any case, it is rather unlikely that any fragmentation scenario in the hot Big Bang picture could produce fragments as small as this. One might, on the other hand, envisage the fragments being as small as *jupiters* (i.e. objects in the mass range $M < 0.08 \, M_\odot$ which are too small to ignite their nuclear fuel). We will see later that such objects could only be detectable by their gravitational lensing effects.

There would be a better chance of detecting Population III objects which derive from nuclear-burning stars. Stars smaller than $1 \, M_\odot$ would still be burning but they would have to be at least as small as $0.1 \, M_\odot$ in order to have a mass-to-light ratio large enough to explain any of the dark matter problems. Such *M-dwarfs* (as they are termed) could in principle contribute to the dark matter but they would have to reside in the narrow mass range 0.08–$0.1 \, M_\odot$. Population III stars larger than $1 \, M_\odot$ would no longer exist but they could still have produced dark remnants. For example, *white dwarfs* and *neutron stars* could derive from stars in the mass ranges 1–$4 \, M_\odot$ and 8–$100 \, M_\odot$, respectively. Stars in the intermediate mass range (4–$8 \, M_\odot$) are thought to explode during carbon-burning. Sufficiently large stars in the range below $100 \, M_\odot$ could evolve to *black holes*. It is not certain at what mass the boundary between neutron star and black hole remnants occurs but it is probably around $20 \, M_\odot$.

Table 9.2. *Stellar evolution. This indicates the mass range, fate and type of remnant associated with the various possible Population* III *objects. The mass ranges are only approximate and there are uncertainties in some of the fates. SMO stands for 'Supermassive Object', VMO for 'Very Massive Object', MO for 'Massive Object', and LMO for 'Low Mass Object'*

Type	Mass ($M_\odot$)	Fate	Remnant
SMO	$> 10^5$	Collapse due to relativistic instability before H-burning	Black hole
VMO	$200{-}10^5$	Collapse during O-burning due to pair instability	Black hole
	$100{-}200$	Explode during O-burning due to pair instability	None
MO	$8{-}100$	Core collapse plus envelope ejection after Ni-burning	Neutron star or black hole
	$4{-}8$	Explode during C-burning	None
	$1{-}4$	Collapse after H/He-burning	White dwarf
LMO	$0.08{-}1$	Still undergoing H-burning	M-dwarf
	< 0.08	Never ignite nuclear fuel	White dwarf or jupiter

Although stars larger than $100\,M_\odot$ are rather rare at the present epoch, they may have formed more prolifically in the Population III context. Such *Very Massive Objects* could also be efficient generators of dark matter; indeed that is one of the prime reasons for invoking them in the Population III context. Those in the range $100{-}200\,M_\odot$ would explode during oxygen-burning but those larger than $200\,M_\odot$ could collapse to black holes completely, without any prior mass ejection. A still more exotic possibility is that the Population III objects may have been larger than $10^5\,M_\odot$. Such *Supermassive Objects* would have collapsed to black holes even before burning their nuclear fuel as a result of relativistic instabilities.

Table 9.2 summarizes the evolution of these different sorts of stars in more detail. Depending on their initial mass, they are conveniently classified as 'Low Mass', 'Massive', 'Very Massive' or 'Supermassive'. The table makes it clear that there is a wide range of masses for which one expects to be left with dark remnants, so the suggestion that 'Population III' objects can make dark matter is not implausible providing enough of them can form.

4   Constraints on the dark matter

The considerations of the last section show that there are a large number of ways in which dark matter could arise. Some of the candidates might be regarded as rather exotic but none of them can be rejected at the outset. The problem then is to narrow down the range of possibilities. After all, the candidates listed in Table 9.1 span a range of masses from 10^{-5} eV to $10^9 M_\odot$, corresponding to 80 mass decades. It is rather perturbing that most of the Universe should be in a form which is so loosely constrained!

We will first discuss the constraints on the 'Population III' candidates. One of the most important constraints is associated with *cosmological nucleo-synthesis*. In the standard hot Big Bang picture, the neutron-proton ratio freezes out at a temperature $T_F \sim 1$ MeV (i.e. at a time $t_F \sim 1$ s), when the rate for the weak interactions involving these particles falls below the cosmological expansion rate. At this point the ratio has a value of about $\frac{1}{8}$. Since all neutrons (except the small fraction which are lost through β-decay) burn first into deuterium and then into helium at about 100 s, the resulting helium abundance is roughly 25% by mass. There are also small residual abundances of deuterium, helium-3, and lithium-7. These depend very sensitively on the total baryon density, but it is a remarkable triumph of the standard model that the predicted abundances of all these elements are consistent with observation providing the baryon density is about 10% of the critical value. Thus the dark matter in galactic halos (and conceivably clusters) could be of baryonic origin, but a critical density certainly could not be unless one sacrifices the usual Big Bang picture altogether. We infer that Population III remnants could not close the Universe. However, this conclusion would not pertain for inos since they do not participate in cosmological nucleosynthesis. Nor would it pertain for primordial black holes that formed before the neutron-proton freeze-out time; such holes would necessarily be smaller than $10^6 M_\odot$.

An important constraint on Population III stars which have already burnt out is associated with *metallicity production*. The existence of 'Population I' stars with fractional mass in metals as low as 10^{-3} excludes any of the dark matter problems being solved by stars which produce an appreciable heavy element yield. In particular, this excludes neutron stars as an explanation of any of the dark matter problems. For neutron stars can only derive from stellar precursors which end their lives as supernovae and such stars return at least 10% of their original mass to the background medium as heavy elements. This implies that at most 1% of the background gas could be turned into such precursors. The metallicity limit does not apply for white dwarfs since these derive from stars which return helium rather than metals to the background medium. However, even in this case, the fact that only a fraction of the mass of the precursor ends up in the remnant would allow white dwarfs to explain only the local dark matter problem.

Since stars produce radiation during their nuclear-burning phase, another constraint on Population III stars which have already burnt out can be inferred from *background light limits*. However, these limits are very dependent on the epoch at which the stars burn. For example, the background light from a population of very massive stars can generally be pushed below the observational upper limit by having the stars burn early enough. Nevertheless, one can still restrict the density of stars in the mass range $1-10\,M_{\odot}$ since these have such a long lifetime that they necessarily burn until a rather late epoch.

A useful constraint on Population III candidates which are still burning their nuclear fuel comes from *source count limits*. We have seen that stars with mass around $0.1\,M_{\odot}$ might in principle have a sufficiently high mass-to-light ratio to explain any of the dark matter problems. However, such stars could still be detectable as high velocity infrared sources and infrared searches already indicate that their number density near the Sun can be at most $0.01\,\mathrm{pc}^{-3}$. This is a hundred times too small to explain the local dark matter and ten times too small to explain the halo problem. A similar conclusion is indicated by infrared observations of other spiral galaxies; these suggest that the mass of the halo objects must be less than $0.08\,M_{\odot}$, which may preclude any nuclear-burning stars at all. Such observations do not exclude jupiters, of course, since jupiters are dim even in the infrared.

Another sort of constraint is associated with *gravitational lensing* effects. This constraint arises because relativity theory predicts that light should bend as it passes through a gravitational field, thereby producing image-doubling. If one has a population of objects with mass M, then the probability that one of them will lie close enough to the line of sight of a quasar to image-double it is just of order the density of the objects in units of the critical density; the separation between the images is $10^{-6}\,(M/M_{\odot})^{1/2}$ arcsec. Thus the 'VLA' radio telescope – with a resolution of 0.1 arcsec – could search for lenses as small as $10^{10}\,M_{\odot}$ (indeed it has already found lensing galaxies), and the 'VLBI' technique – with a resolution of 10^{-3} arcsec – could search for ones as small as $10^{6}\,M_{\odot}$.

This effect is important only for very large Population III objects. However, a galaxy itself can act as a lens and, if one is suitably positioned to image-double a quasar, then it can be shown that there is also a high probability that an individual halo object within the galaxy will traverse the line of sight of one of the images. This will give appreciable intensity fluctuations in one but not both images. This effect would be observable for stars larger than $10^{-4}\,M_{\odot}$ but the timescale of the fluctuations, being of order

$40(M/\mathrm{M_\odot})^{1/2}$ yr, would only be detectable over a reasonable period for $M < 0.1\,\mathrm{M_\odot}$ (i.e. for jupiters). However, yet another kind of lensing effect could permit the detection of objects with $M > 0.1\,\mathrm{M_\odot}$. This is because such objects could modify the ratio of the line flux to continuum flux of the quasar. The fluxes are affected differently because they come from regions of different size: unless M exceeds $10^5\,\mathrm{M_\odot}$, the line radiation comes from a region which is too large to be lensed, whereas the continuum radiation comes from a much smaller region which can be. This already excludes a critical density of objects with $0.1 < M/\mathrm{M_\odot} < 10^5$, though not necessarily the tenth critical density required for halos. Supermassive black holes are not excluded by this effect because, if M exceeds $10^5\,\mathrm{M_\odot}$, even the continuum region cannot be lensed. However, we have seen that such holes might be detected by their direct image-doubling anyway. Thus gravitational lensing constrains the Population III mass spectrum over nearly the entire range above $10^{-4}\,\mathrm{M_\odot}$.

Finally a variety of *dynamical constraints* restrict the masses of any dark baryonic objects. For example, the survival of binaries in the galactic disk requires that the objects which comprise the local dark matter be smaller than $2\,\mathrm{M_\odot}$. In particular, this excludes black holes. The requirement that the disk should not be puffed up too much by the heating effect of traversing halo objects implies that the halo objects must be no larger than $10^6\,\mathrm{M_\odot}$. Even if the dark matter in clusters is different from the halo dark matter, the absence of unexplained tidal distortions of visible galaxies in (for example) the Virgo cluster implies that the dark objects must still be smaller than $10^9\,\mathrm{M_\odot}$. The fact that any dark matter which makes up the critical density must avoid clustering like galaxies implies that it must have a velocity dispersion of at least 10^3 km/s (probably excluding any Population III candidates) unless one invokes biased galaxy formation.

All of the constraints discussed so far pertain to Population III candidates for the dark matter. Most of the constraints on the elementary particle candidates are dynamical in nature and depend on the fact that we need the particles not only to provide an appreciable cosmological density but also to cluster on various scales. For example, we can almost certainly exclude the disk dark matter being inos because they would not be able to cool enough to form a disk configuration. We can also exclude certain types of inos from providing the halo dark matter. This is because quantum effects preclude particles of a given momentum getting too close together, so the clustering requirement places a lower limit on the particle's mass. For example, if hot particles like neutrinos are to provide the dark matter in galactic halos, then one can argue that their mass must exceed 20 eV. But in this case, their density is 0.5 of the critical value, which is marginally excluded.

Such considerations do not exclude hot particles from comprising the dark matter in clusters (since both the density and clustering requirements would be consistent with a particle mass of 10 eV). However, even this possibility may be implausible because the first bound objects tend to form too late. The reason is that the 'free-streaming' of inos in the period when they are moving relativistically will erase any fluctuations in their number density on scales less than the horizon size when they go non-relativistic. (The horizon size is just the distance travelled by light since the big bang, approximately the age of the Universe times the speed of light.) This eliminates fluctuations on mass-scales below $\sim 10^{15} M_\odot$. This scale is very large, so the first structures to form will be giant 'pancakes' of supercluster scale. However, since very large scales necessarily bind very late, it turns out to be very difficult to explain how the observed large-scale structure could have evolved by the present epoch.

Rather similar criticisms apply to warm particles like the gravitino, although the problems are then less extreme. For this reason many cosmologists now regard some cold particle like the axion or the photino as the most plausible candidate for providing a critical density. Although one would naïvely assume that cold particles would cluster too easily to provide a uniform background, this objection can be circumvented if one invokes biased galaxy formation.

5   Conclusion

The various constraints discussed above are brought together in Table 9.3, which indicates the candidates which could explain each of the four dark matter problems. The shaded regions in this figure are excluded by at least one of the arguments given above. The prime message of Table 9.3 is that one should not expect any single dark component to resolve all four dark matter problems. This should be of little surprise in view of the preponderance of dark objects in the Universe. On the other hand, the figure does give some indications of what the best solutions might be: (i) the best candidate for the local dark matter would seem to be white dwarfs or jupiters; (ii) a possible solution for the halo dark matter would seem to be the black hole remnants of Very Massive Objects, though one cannot exclude jupiters, primordial black holes, or warm or cold inos; (iii) the dark matter in clusters would need to be primordial black holes or inos if one adopts the cosmological nucleosynthesis limit in its strongest form but the other halo candidates would be viable if one adopts it in a weaker form; (iv) the closure dark matter could only be inos or primordial black holes.

Our analysis does not provide a unique answer to the four dark matter problems but it at least narrows down the range of possibilities. Lacking a unique answer, cosmologists tend to assess the likelihood of the various candidates according to their individual prejudices. Thus, as indicated in

Table 9.3. *Constraints on the dark matter. This summarizes the constraints on the form of the dark matter discussed in Section 9.4. The four dark matter problems are discussed in Section 9.2 and the various solutions in Section 9.3. PBH stands for 'Primordial Black Hole', NS for 'Neutron Star', WD for 'White Dwarf'; the other abbreviations are indicated in the caption for Table 9.2. The shaded regions are excluded.*

		LOCAL	HALO	CLUSTER	CLOSURE
POPULATION III	SMO				
	VMO				
	MO				
	NS				
	WD				
	LMO				
INOS	PBH				
	COLD				
	WARM				
	HOT				

MASS ↑

Section 1, particle physicists tend to prefer ino solutions, while astrophysicists tend to prefer Population III solutions. In a spirit of compromise, however, it is perhaps worth stressing that both ino and Population III solutions may be relevant; inos could provide the closure density and perhaps the dark mass in clusters, while black holes may provide the dark matter in halos. It is in this spirit of compromise that the third picture in Fig. 9.2 is offered!

10

Geometry and the Universe

C.V. VISHVESHWARA

The little holy book

At the age of twelve, Albert Einstein came to possess a short book on Euclidean geometry. Decades later, he was to recall in his *Autobiographical Notes* the profound influence this little 'holy geometry booklet' had exercised on him.

'Here were assertions ... which – though by no means evident – could nevertheless be proved with such certainty that any doubt appeared to be out of the question. This lucidity and certainty made an indescribable impression on me.'

Even when so young, Einstein had already grasped the potent characteristics of geometry: the lucidity of deductive reasoning and the certainty of proof of assertions 'by no means evident' starting from a handful of axioms. It was precisely these aspects of geometry that had made it the basis of all mathematics for more than two thousand years. At the same time, the brilliant intuitive powers of the Pythagoreans had recognised the indispensable need of mathematics for the true understanding of nature, a fact to be amply confirmed by centuries of scientific enquiry that followed. It was inevitable then that geometry, the corner stone of mathematical thought and methodology, should be at the heart of the description of the cosmos. Because of this, Galileo was led to make his famous statement,

'Philosophy is written in that great book which ever lies before our eyes – I mean the universe – but we cannot understand it if we do not first learn the language and grasp the symbols in which it is written. The book is written in the mathematical language, and the symbols are triangles, circles and other geometrical figures, without whose help it is impossible to comprehend a single word of it; without which one wanders in vain through a dark labyrinth.'

The symbols were of course those of Euclidean geometry. Everywhere in the universe, at all scales of size and complexity, one could discern the ordaining rules of this geometry at work – from the lovely, intricate forms of the most minute organisms to the precise planetary orbits that framed the heavens. The profundity of the concepts underlying those rules has always deeply influenced human thought at the highest and the most abstract level. On the other hand, the element of the miraculous inherent to the strange and surprising properties of certain geometrical figures has been an invitation to mysticism, sometimes ensnaring even the most rational minds. All this – the rules, the ideas and the magic – that had been put together to depict the universe during a period of more than two millennia could possibly have been derived from the 'holy geometry booklet'. But, the new cosmos to be discovered by Einstein would far transcend the system of knowledge represented by the little book. It would be based on non-Euclidean geometry with all its strange possibilities, opening up new vistas in cosmology and revealing phenomena undreamt of in the old philosophy. Today as the quest for new theories and deeper understanding of our universe continues, geometry seems still to dominate our thinking, however novel its new incarnations might be and however different the language it might be couched in. But, let us not forget that deep beneath the unfinished edifice the foundations one finds belong to Euclidean geometry, a small sample of which must have been enshrined in the little holy book.

Images of reality

Because of its strong appeal to our natural instinct for visualisation, the very word 'geometry' evokes mental pictures of curves, surfaces and volumes. The visual aspect of geometry has often played a significant role in many different contexts. For instance, the Pythagoreans to whom numbers were the ultimate constituents of the universe, pictured their characteristic arrangements into geometric patterns. These patterns provided a framework for their theory of numbers. They could thus think of triangular numbers, oblong numbers, pentagonal numbers and so on. But only whole numbers could be accommodated within this system. According to legend, Hippasus of Metapontum was thrown into the sea by the Pythagoreans for the sin of having discovered irrational numbers which did not legitimately belong to their universe.

Curves and surfaces are the basic ingredients with which any structure is designed and built. The architecture of the universe itself is no exception to this. From the very early times, the visualisation of the universal geometry has been one of the natural aims of cosmology. The trajectories traced by the celestial bodies are the curves that delineate this cosmic design. The Babylonians, who maintained records of planetary movements and had at their disposal adequate mathematical techniques to predict their positions, failed however to visualise the planetary orbits. The evolution of a geometric picture of the universe had therefore to wait for the early Greek mathematicians and philosophers to appear on the scene. Our present conception of the universe, as we shall see, is based on the curved four-dimensional geometry of spacetime. It is impossible for a normal mind to visualise a curved three-dimensional space let alone a four-dimensional one. Nevertheless, the modern geometer studies two-dimensional slices of such a spacetime and tries to imagine a composite picture of the universe built out of these partial views.

Concrete, quantitative connections between geometry and the large-scale structure of the universe are established through astronomical measurements. It is not surprising therefore that offshoots of geometry owed their development to the needs of astronomy. Trigonometry, for instance, was built up and refined in order to facilitate the prediction of the paths and the positions of celestial bodies. The thirteen books of Ptolemy's *Almagest* compile both astronomy and trigonometry, the two constantly going hand in hand. Trigonometrical theorems are proved only when and where they are necessary for astronomical application. In a like manner, much of the ancient Indian mathematics, including trigonometry, was motivated by the need for developing quantitative astronomy. As a matter of fact mathematics was almost invariably included as part of astronomical treatises.

The real power of geometry lay in its method of abstraction, clear conceptualization and above all in its infallible deductive reasoning. A few axioms, taken to be self-evident, were sufficient to generate an extraordinary number of rigorous results whose truth was beyond doubt. It is indeed a staggering fact that Euclid's *Elements* contains no less than 467 propositions and Appolonius' monumental work *Conic Sections* 487, all derived from the ten axioms of Euclid. It is this inherent logical rigor as well as its structural beauty that made geometry the basis and model for other branches of mathematics. The opinion that the latter could draw strength from geometry, but could never rival it prevailed even in the eighteenth century. Algebra and calculus were considered inferior to geometry. Newton, in his unbounded admiration for Greek geometry considered algebra 'the analysis of the bunglers of mathematics' and Leibnitz for his part maintained that 'geometers could demonstrate in a few words what is very lengthy in the calculus'. Naturally, for centuries, the supreme task of providing a mathematical description of the universe was reserved for geometry. In fact, the very purpose of geometry was identified with that of understanding nature.

In the quest for the comprehension of the universe, the early Greek ideas – especially those embodied by Plato's philosophy – were the torchbearers that lighted the paths laid out by geometry. Not only did those ideas exert a strong influence on the development of cosmology, but we can often recognise some of them, at least in spirit, infused into the tenets of modern science. Let us then take a quick look at a few aspects of the Greek view concerning geometry and the universe.

First of all the study of geometry was a preparation required of every philosopher. It helped in the clear perception of abstract ideas and the formulation of concepts transcending physical appearances. Plato wrote in his *Republic* that though geometers 'make use of visible forms and reason about them, they are thinking not of these, but of ideas they resemble ... but they are really seeking to behold the things themselves, which can be seen only with the eye of the mind'. Further, geometry forced 'the soul towards the region of that beatific reality which it must by all means behold'.

This reality, as it manifested itself in the universe, consisted of different levels. The world of experience accessible to sensory perception belonged to a lower stratum of reality than the one that could be comprehended only by pure intellect. The former was subject to constant change, while the latter was immutable and eternal. And 'geometry is knowledge of the eternally existent'. This knowledge totally transcended the world of experience. Physical phenomena were important not for their own sake, but only in as much as they could transport one to higher reality. Take astronomy for example. Along with arithmetic and music it was closely associated with geometry. Nevertheless, its concerns were confined merely to the visible world. It forced the eyes and not the mind to look upwards and failed to reveal 'the unseen reality'. In other words, the paths of the planets, 'those intricate traceries in the sky are no doubt the loveliest and most perfect of material things, but still part of the visible world, and therefore fall short of the true realities'. The 'true realities' were composed of pure numbers and perfect geometrical figures. They could never be reproduced by material bodies. The only useful purpose of astronomy was to bring forth problems that could stimulate the study of geometry. Similar was the situation in regard to music, the kindred discipline of astronomy. The practice of music by itself did not reveal the truth of harmonies embodied by numbers and their ratios. The 'worthy musicians who tease and torture the strings, racking them on the pegs' – much the same way the slaves were tortured in order to force confessions out of them – could hardly unravel the mystery of numbers. Nor could the astronomers merely by the study of the heavens arrive at the truth underlying the cosmic structure. The universe was identified with the world of ideas. And the truth underlying it transcended all sensory experience. To this world belonged the abstract geometrical concepts.

Plato's ideas were modified by Aristotle. According to his philosophy, abstract mathematical concepts were derived from the world of experience and in turn they represented the observable physical phenomena. Yet, there was no denying that only the intellect could arrive at the ultimate reality. The study of the universe was in essence part of this heroic quest for truth. Geometry, 'the knowledge of the eternally existent', provided the natural model for the structure of the universe.

The unshakable superstructure of reality rested firmly on the few axioms of geometry whose truth was self-evident. This was the sturdy framework on which the entire universe hung together. But, how could one tell that the axioms were true beyond a shadow of doubt? According to Plato, this was indeed possible because the human mind, by virtue of traits carried over from previous birth, could distinguish the true from the false with absolute certainty. Nevertheless, strangely enough, starting from the time of Euclid himself doubts in regard to the self-evident nature of one of the axioms lingered and grew over the centuries culminating finally in the creation of non-Euclidean geometry. The new geometry would herald the modern description of the universe as accepted today. But doubts can arise only when there are firm assertions. It is an astonishing fact that such assertions concerning the truth of geometrical postulates were made with absolute clarity and conviction during the very birth of mathematics.

Since the mind was the supreme agent that could comprehend the real universe, whatever appealed to it had to be accepted as the truth. Beauty and simplicity, two qualities for which the mind had an affinity, were the natural attributes demanded by the correct model of the universe. The planetary orbits had therefore to be circles, the perfect figures that can be drawn on a plane. Further, the planets must move with constant speeds. This made their paths perfect in both space and time: all points were identical to one another and every moment of time was the same as the other. The Pythagorean cosmology consisted of crystal spheres that revolved uniformly around the earth, carrying the sun, the moon and the planets imbedded in them, and filled the heavens with the silent melodies of their celestial harmony. God, being the creator and the ultimate personification of this perfection, had to be spherical in shape too. As early as in the 6th century B.C., Xenophanes had declared that the 'universal homogeneity of God implies that he has the shape of a sphere ... And because he is uniformly the same and round as a ball, thus he is neither limited nor unlimited, neither in rest nor in motion'.

Surprisingly, many tenets of the Greek philosophy of nature and its mathematical representation still form the basis of modern scientific enquiry. Beauty and simplicity continue their role as the precious touchstone of a scientific theory. Hermann Weyl is quoted to have remarked that he preferred a theory that was inherently beautiful, though seemingly wrong, to the one that was correct but cumbersome. We certainly recognise the need for involving experiment and observation as the starting point or the final test of a mathematical model portraying some aspect of nature. Yet, while building up a mathematical theory from basic principles, exploring its implications and working out its consequences, we do live on a separate plane of reality sometimes far removed, however temporarily, from the world of direct experience.

The geometrization of the physical world continued with vigour over the centuries that followed the Greeks. In the 17th century René Descartes invented coordinate geometry which proved to be a powerful tool in the investigation of curves and surfaces. The description of nature involved, for instance, the trajectories of material particles which found their representation in mathematical curves. The shapes of objects one encountered could be identified with mathematically defined geometrical figures. Geometry thus appeared to be at the very hub of the physical world. Descartes himself was to remark therefore that he 'neither admits nor hopes for any principles in physics other than those which are in geometry or in abstract mathematics, because thus all phenomena of nature are explained and some demonstrations of them then can be given'. Similar were the views of Leibnitz who held that 'the whole matter is reduced to pure geometry, which is the one aim of physics and mechanics'. The programme of geometrization of physics was greatly strengthened by other mathematical physicists – for example William Hamilton whose approach to mechanics was predominantly geometrical. All this seemed to lend support to Plato's aphorism, 'God eternally geometrizes'. Not that the unchallenged importance assigned to geometry did not vacillate and wane during the course of time. While Jacobi was to counter Plato by his own assertion that 'God ever arithmetizes', James Joseph Sylvester went so far as to remark: 'Geometry may sometimes appear to take the lead over analysis but in fact precedes it only as a servant goes before the master to clear the path and light him on his way'. Other branches of mathematics advanced independently of geometry and their value to physics was recognised in due course. Nevertheless, geometry was destined to make a dramatic comeback not in its Euclidean form but as a startling departure from it.

Of magic and mysticism

'Metaphysics, or the attempt to conceive the world as a whole by means of thought, has been developed, from the first, by the union and conflict of two very different human impulses, the one urging men towards mysticism, the other urging them towards science.' So wrote Bertrand Russell. Often, geometry has been the bridge between these two impulses. The invariant order of geometric figures and the strange inter-relations among their constituent elements could not fail to cast a spell of enchantment on anyone who studied them. Here was a system of precise knowledge governed by unerring rules endowed with the touch of the miraculous. One who mastered this science could surely possess the key to the doors that opened upon cosmic knowledge. Geometry was the magic nexus between human comprehension and the divine will. It was the language by which the mortals could communicate with and propitiate the gods. Might it not even be the means to control the cosmic forces that ordained man's fate? This then was the essence of the mysticism that was inspired by geometry. In turn, that mysticism, when professed by keen minds, paid its debt to geometry by nurturing it further.

The earliest Indian writings on geometry, the *Śulvasūtras* or 'the rules of the cord' – so-named because cords were employed for the measurement of length – are directly linked to the construction of sacrificial altars. The sacrifices were intended to please the gods who would grant the wishes of the sacrificer. A falcon shaped altar was required if one desired heaven. The sacrificer could thus become symbolically a falcon himself, the best flyer in the world, and soar to the heavenly abode. An altar in the shape of a chariot wheel represented thunderbolt which the sacrificer could hurl on his enemies and destroy them. These and other altars of different shapes had one vital factor in common: their areas were equal to that of a canonical square altar. According to the ancient Indian philosophy and theology, the square embodies the perfect, fundamental form. Both the construction of altars and the architecture of temples are based on the square as the starting point. All other shapes play a secondary role. The circle, for example, represents a state of flux, the expanding energy from its centre evolving its shape, ultimately culminating in the square. The square, on the other hand, symbolizes order,

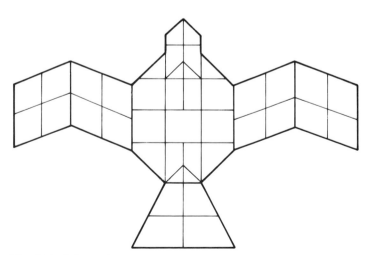

Fig. 10.1. Schematic representation of *Vakrapakshasyenachit*, the falcon shaped altar.

stability and the final state of evolving life. It is perfection beyond life and death. The essence of the square would be retained by any *mandala*, or a closed polygon, as long as the area was kept unaltered. The main thrust of the *Sulvasutras*, therefore, does not lie in the superstitious beliefs they are supposed to serve, but in the geometrical operations needed for the planning of altars of different shapes, but of the same area as that of a prescribed square.

It is evident from the *sutras* that the theorem of Pythagoras was well known to the master geometers of this era when western mathematics had not yet seeped into the Indian soil. They made abundant practical use of the theorem in erecting altars with complex structures. The problems involved in this task included that of 'squaring of the circle' – or constructing a circle with area equal to that of a given square – a problem which 'has since haunted so many unquiet minds'.

Was the motive behind all the mathematical analysis underlying the *sutras*, as Thibaut puts it, 'simply the earnest desire to render sacrifice in all its particulars acceptable to the gods, and to deserve the boons which the gods confer in return upon the faithful and conscientious worshipper'? In other words, were the requirements acceptable to the gods first prescribed rigidly and then the geometrical rules and techniques discovered and evolved to satisfy them? Or was it the other way round? Once the miraculous geometrical relations dictating the properties of curves and figures were discovered, was it not natural to borrow the mysterious powers behind them in order to manipulate the supernatural world of gods and demons? One can only speculate. But, there is no doubt that when the line demarcating science and mysticism was ill-defined, the two elements could not but strongly influence each other.

The Pythagoreans believed that the Creator or the Ordering One had structured the Cosmos – the word originally meant Order – as a harmonious whole in concordance with the symmetries and correlations intrinsic to the numbers and their geometrical representations. Stemming from this conception of the universe grew the complex mysticism of numbers and geometrical figures. Specific characteristics were attributed to them and hidden meanings read in their interrelations. For instance, five was considered to be the most important number, the symbol of health and harmony; it was the emblem of love, being the offspring of the union between the first female number two and the first male number three. Its geometric counterpart, the pentagram, exhibited so many intriguing, magical properties that it was chosen as the sign of brotherhood among the Pythagoreans. The secret knowledge originating from the Pythogoreans cascaded down the centuries along diverse channels, such as the cults of the Kabbala, Freemasonry and Rosicrucianism. But never did the original geometric mysticism exercise such a profound and far-reaching influence on an individual as in the case of Kepler during his early years that witnessed his quasi-scientific fantasies slowly transmuted into the pure gold of his three laws of planetary motion.

The inspiration that fashioned Kepler's ideas and fuelled his relentless explorations had its roots deep within the Pythagorean and Platonic geometric mysticism. He expounded the divine status of geometry in his *Harmonices Mundi*, the Harmony of the World, 'Why waste words? Geometry existed before the Creation, is co-eternal with the mind of God, *is* God *himself* (What exists in God that is not God himself?); geometry provided God with a model for the Creation and was implanted into man, together with God's own likeness – and not merely conveyed to his mind through the eyes.'

God picked as his building blocks of the universe the Platonic solids because of their perfect symmetry. It is an incredible fact, but true, that, although there can be an infinite number of symmetric two-dimensional figures, there exist five – and only five – such solids in three dimensions. In each instance, the solid is bound by faces which are themselves symmetric two-dimensional geometric figures of a particular kind. These are the tetrahedron (made of four equilateral triangles), the cube, the octahedron (eight equilateral triangles), the dodecahedron (twelve pentagons) and the

Fig. 10.2. The five regular Platonic Solids. From Arthur Koestler, *The Watershed*. Anchor Books, Doubleday and Co. Inc., Garden City, New York 1960.

icosahedron (twenty equilateral triangles). Each of these corresponded with a basic element of nature, for example, the tetrahedron with fire, the cube with earth and so on. The dodecahedron, being composed of the magical pentagons, occupied the highest place among the constituents of the universe. While geometry admitted this unique set of precisely five perfect solids, was it a mere accident that nature had decreed the existence of six planets with exactly five intervals in between? Surely, this was a crucial clue to the mystery of the cosmic structure. This explained why there were six planets, no more no less. Would not the five solids decide the sizes of the planetary orbits and the order in which they had been arranged as well? The orbits could be drawn as circles on spheres either circumscribing the Platonic solids or inscribed within them. When chosen properly, the five solids seemed to fit roughly the known orbits of the six planets. Kepler was to remark on this revelation in 'the preface to the reader' of his *Mysterium Cosmographicum*, 'I saw one symmetric solid after the other fit in so precisely between the appropriate

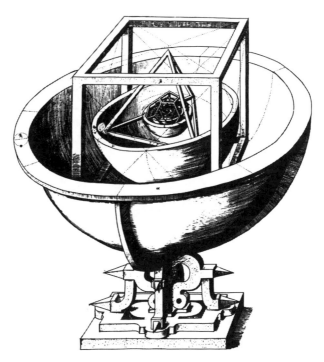

Fig. 10.3. Kepler's scheme for the planetary orbits based on the Platonic solids. From Arthur Koestler, *The Watershed*, Anchor Books, Doubleday and Co. Inc., Garden City, New York. The original is from Kepler's *Mysterium Cosmographicum*.

orbits that if a peasant were to ask you on what kind of hook the heavens are fastened so that they don't fall down, it will be easy for thee to answer him. Farewell!'

This was only the beginning and not the end. Had Kepler ceased his probing at this stage, satisfied by the fruition of his fantasies, his name might not have found a place even in the footnotes to astronomy and cosmology. But, his scientific objectivity drove him to test and justify his model against the available astronomical data. In this venture, he was obliged to replace his two-dimensional spheres by spherical shells in order to accommodate deviations of the orbits from perfect circles – the first step towards the realization of elliptical orbits. Obsessed with the desire to prove the validity of his model, not only did he criticise Copernicus for his inexactitude, but he even dared to accuse him of deliberate cheating. Nevertheless, years of excruciatingly painstaking research finally converged on to the three famous laws of planetary motion. With these laws firmly established, the artificial scaffolding of the universe Kepler had erected could be dismantled and the true order of the cosmos revealed. The planetary orbits were asymmetric ellipses and not the ideal circles of the Greeks. On the other hand, with Kepler's work, Apollonius' *Conic Sections*, a purely intellectual creation, found its niche within the realm of cosmic reality. Plato in his *Timaeus* had described how God had created the planets as the regular keepers of time. Kepler had now found the exact law that regulated their divine duty. In addition to all this, it was Kepler who introduced, however vaguely and hesitantly, the idea of a force – or 'soul' as he termed it – emanating from the sun and impelling the planets in their heavenly courses. Starting from the nebulous mysticism of an antique era, Kepler had unveiled the true face of the cosmos for contemplation by other giants like Newton. God the Geometer would soon become God the Watchmaker maintaining his Creation in perfect repair. Still, on the terrestrial scale it was geometry that created, by the skilful use of delicious curves and surfaces, an enchanting garden of delights.

The garden of earthly designs

Descending from the heavens to the earth, one can discover geometry at work in all its glory. Myriad handiworks of nature display intricate patterns with perfect order and symmetry. Forces acting from within both organic and inorganic systems, often in combination with environmental factors, decide the different classes of regular shapes these systems will evolve into. Curves and surfaces, conceived and studied purely from mathematical interest, find their expression in a variety of exquisite natural specimens. The symmetric formation of crystals into beautiful geometric shapes is a common example. Organic growth by accumulation of matter as in the case of horns,

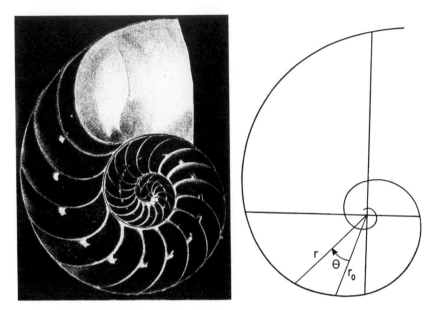

Fig. 10.4. The equiangular spiral structure of the *Nautilus* shell. The
mathematical curve is given by the equation $r = r_0 a^\theta$. (Photograph
from D'Arcy Thompson, *On Growth and Form*, Cambridge University Press, 1942.)

tusks, and especially shells follows along the widely studied equiangular or
logarithmic spiral. Here the requirement imposed by nature is that each
successive increment of growth be self-similar to its predecessor in shape and
relative positioning except that it is a magnified version of the latter. And the
mathematical curve – the logarithmic spiral – fulfils the requirement to
perfection. Examples such as these and detailed studies on them can fill
volumes.

A basic law regulating the formation of specific geometric structures is the
extremization principle. That is, some particular quantity is usually minimized – and sometimes maximized – in natural processes thereby accounting
for the behaviour of the system under consideration. For instance, way back
in the 1st century A.D., Heron had pointed out that the law of reflection of
light implied that the light ray took the shortest path in its passage between
two points, which demonstrated nature's familiarity with geometry. Fermat's
principle generalized this to the minimization of time of passage of light
between two points taking into account different media in between. The
minimization principle evolved and developed into a powerful mathematical
tool capable of explaining not only the geometric patterns of static configurations, but also the dynamic behaviour of mechanical systems. Moreover, the

principle, which clearly exhibited the economy of nature, was elevated to the status of an all-embracing doctrine with philosophical and theological overtones. To Pierre Maupertuis, the founder of the principle of least action, it was a proof of the very existence of God. Leonard Euler, who made outstanding contributions to the development of the principle, wrote: 'Because the shape of the whole universe is most perfect and in fact, designed by the wisest creator, nothing in all of the world will occur in which no maximum or minimum rule is shining forth.'

A classic example of geometric structure, recognised by mathematicians of all ages, in which the 'minimum rule is shining forth' brightly is that of the honeycomb. While the lovely hexagonal symmetry of the cross-section of each cell is the result of its pushing against its neighbours on all sides, the task of explaining the shape of the closed end of the cell constitutes a challenging problem in mathematics. Surprisingly, it was the great Kepler who discovered that the end planes formed part of a rhombic dodecahedron. The exact measurements were made by the famous astronomer Cassini's nephew Maraldi, himself an astronomer. The theoretical explanation for the geometric shape of the cell was offered in the early eighteenth century by the Swiss mathematician Samuel Koenig utilizing the minimum principle: this was the configuration which required the least quantity of material for its construction. After all, Pappus of Alexandria, as early as in the 3rd century, had conjectured that the bees were endowed with 'a certain geometrical forethought'. Now, Koenig claimed that the problem the bees had solved was beyond the reach of old geometry. They must therefore have used the recently developed methods of calculus. When called upon to pass his judgment on this case as the Secretary of the French Academy, Fontenelle, while denying intelligence to the bees, suggested however that they were employing highest mathematics by divine guidance and command. As a postscript to this episode, we may note that the most modern and accurate solution to this problem is slightly different from the actual construction followed by the bees. The difference is about 0.35% and the bees could be wasteful by this amount. In all likelihood, Fontenelle would have been delighted to know this fact!

Soap bubbles provide another example of minimum principle in the geometry of nature. Here the work done in creating the bubble against the force of surface tension is the quantity that is minimized. Since this work is proportional to the area of the liquid film, the bubble assumes the shape with the least area. Of course, in the case of a free bubble this is simply a sphere. On the other hand, bubbles can also be produced by dipping wire frames in soap solution. One is then confronted with Plateau's problem of obtaining surfaces of minimal area with specified boundaries. The soap bubbles assume a variety of unusual forms depending on the shapes of the bounding wire

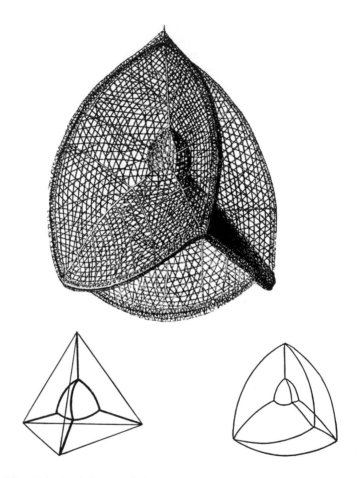

Fig. 10.5. A skeleton of the radiolarian *Callimitra*. It resembles a bubble
within a bubble formed by dipping a tetrahedral wireframe in soap
solution. (Reproduced from D'Arcy Thompson, *On Growth and
Form*, Cambridge University Press 1942.)

frames and are the solutions to Plateau's problem. It is a marvellous fact that
some of these forms are actually reproduced in nature by micro-organisms
like the Radiolaria that radiate sheer beauty and elegance.

Perhaps the simplest geometrical object following the extremum principle
is the geodesic. This is the curve along which the distance between two
neighbouring points is the least. On a plane it is obviously the straightline.
On a sphere, geodesics are given by the great circles like the meridian lines.
The complexity of the geodesics can increase with the shape of the surface on

which they are drawn. Once again, natural realization of these curves can occur in a variety of circumstances. Coils of spiral geodesics peculiar to prolate spheroids can be found in certain spores. The muscular fibres of a hollow organ, such as the stomach or the heart, are arranged along the geodesics of the bounding surfaces of that organ!

The countless examples of exquisite curves and elegant surfaces in nature and their mathematical idealizations are all imbedded in a three-dimensional space which is assumed to be Euclidean. The distances measured in this space and the rules for combining them are taken to be those of Euclidean geometry. Just as a two-dimensional surface immersed in this space can be curved, can the three-dimensional space itself, in which nature functions, be curved as well? Is it meaningful to speak of such curvature and if so, how is it detected? What is the physical agent that could bend three-dimensional space itself? These and other questions, asked and answered, would lead back to a geometric description of the universe at large, radically different from the primitive, mystical models that had been proposed in the past. The charming curves and surfaces inhabiting the Euclidean three space, when generalised to other geometries and higher dimensions, would constitute the structural elements of the new cosmos.

The violets of spring

It is a strange phenomenon that makes a single idea sprout at different places in somewhat different forms but almost at the same time. One needs the right soil and the right season for this to happen. This was indeed the case with the birth of non-Euclidean geometry in the 19th century. In this context, Wolfgang Bolyai, father of János Bolyai one of the founders of non-Euclidean geometry, commented: 'It seems to be true that many things have, as it were, an epoch in which they are discovered in several places simultaneously, just as the violets appear on all sides in springtime.'

While the creation of non-Euclidean geometry shook mathematics to its very foundations, its impact on physics revolutionized our conception of the universe. The developments leading to this dramatic moment in the annals of mathematics followed a tortuous course spanning more than two millennia starting almost from the birth of Euclidean geometry. True, this magnificent edifice rested firmly on just ten axioms considered to be self-evident truths. But, how self-evident was self-evident? There was something disquieting about the fifth postulate, or the parallel axiom, which dealt with the drawing of a straightline through a given point parallel to another straightline. Even Euclid himself had been cautious about the wording of this postulate. The efforts made to improve the rather awkward situation were mainly aimed in two directions. The first of these attempted to replace altogether the fifth postulate by one which was self-evident beyond a shadow of doubt, but equivalent to the original one. A number of outstanding mathematicians were one way or another involved in this enterprise: Ptolemy, Proclus, Nasiruddin of Persia, Legendre and others. The second line of attack consisted in actually trying to prove the fifth postulate as a consequence of the other axioms. Efforts of this kind were made by a number of geometers starting again with Ptolemy and ending more or less with Gerolamo Saccheri. The latter, in the belief that his work had cleared Euclid of all doubts, published a book in 1733 under the title *Euclides ab Omni Naevo Vindicatus* (Euclid Vindicated from All Faults). Although Saccheri's work was a significant milestone in the arduous geometrical journey, his results did not completely settle the question of Euclid's parallel axiom. The enormous amount of energy that had been expended on this frustrating problem without reaching its unequivocal resolution prompted d'Alembert to call it 'the scandal of the elements of geometry'. This was in 1759. In another seventy-five years it would be proved that the fifth postulate was only an assumption, and not a sacred truth and that other geometries as valid as Euclid's, but not conforming to this postulate, could very well exist.

What is this fifth postulate that troubled the deepest intellects and taunted the sharpest talents in mathematics? Simply stated, leaving aside the technical subtleties and philosophical implications, it says that given a straightline and a point not on it, then in the plane determined by the point and the line it is possible to draw one, and only one, straightline through the point parallel to the first line. To all appearances it does seem to be a manifestly evident assertion. Though part of the Euclidean theorems do not call upon this postulate for their proof, taken in its entirety Euclidean geometry does depend on this axiom. On the other hand, as every attempt to prove the postulate or improve it turned out to be futile, the possibility of other geometries independent of this postulate was being entertained with growing seriousness. Gauss, 'the prince of mathematicians', was convinced of this possibility, but refrained from publishing his ideas for fear of being ridiculed by the uninitiated. His work on curved surfaces, however, did incorporate non-Euclidean geometry in two dimensions. This was the first step in the firm establishment of logical self-consistent systems of geometry at variance with the fifth postulate. Yet, till the very dawn of the new geometry, mathematicians despaired over the status of the parallel postulate. Wolfgang Bolyai wrote to his son János, 'You must not attempt this approach to parallels. I know this way to its very end. I have traversed this bottomless night, which extinguished all light and joy of my life. I entreat you, leave the science of parallels alone ... I thought I would sacrifice myself for the sake of the truth. I was ready to become a martyr who would remove the flaw from geometry and return it purified to mankind ... I turned back when I saw that no man can reach the bottom of the night. I turned back unconsoled, pitying myself and all mankind.' But, János had no intention of trying to prove the parallel postulate: his plan was to negate it completely and construct his new geometry. Soon in 1823 he was able to declare to his father: 'Out of nothing I have created a strange new universe!'

János Bolyai and Nikolai Lobatchevsky share in almost equal measures the paternity of non-Euclidean geometry. Their contributions to the creation of the new geometry are extremely similar to each other. Both could build a system of logical geometry, proving theorem after theorem, without incorporating the fifth postulate. In its place, one needed to assume that through a fixed point there were two lines parallel to the first one. As a matter of fact, all the lines lying in between the above two lines happened to be parallel to the first one. This situation can actually be realized in two dimensions on the surface of a *pseudosphere* which looks like an infinitely long trumpet. It is generated by rotating a curve called the *tractrix* around its asymptote. An extremely important factor here is that the 'straightlines' are the geodesics on the surface – the natural generalization of actual straightlines as expected. In addition to the peculiar situation of the parallel lines, one has to give up other common notions such as the angles of a triangle adding up to two right angles. On a pseudosphere the sum of the angles of a triangle formed by three geodesic lines is less than two right angles. On the surface of a regular sphere

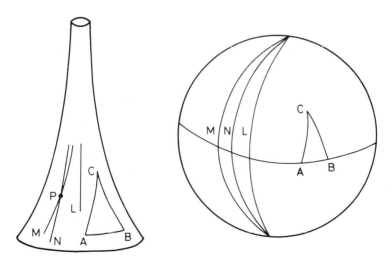

Fig. 10.6. The pseudosphere and the sphere. On the pseudosphere lines *M* and *N* through point *P* approach *L*, but will never intersect it. The sum of the angles of a triangle *ABC* is less than 180°. On the sphere lines such as *L*, *M* and *N* will always meet. The sum of the angles of a triangle *ABC* is greater than 180°.

the sum exceeds two right angles. Of course these are only two-dimensional examples of non-Euclidean properties. But what Bolyai and Lobatchevsky had shown was that such bizarre situations can and do arise even in three-dimensional space governed by the rules of non-Euclidean geometry.

In the saga of the evolving geometric world-picture, perhaps the most significant mathematical development to take place was that of Riemannian geometry in view of its role in modern cosmology. Bernhard Riemann's work encompassed all possible geometries – both Euclidean and non-Euclidean – and provided a powerful tool for the analysis of those geometries. Riemann's approach was fashioned after differential geometry concentrating on the measurement of distance as a starting point rather than on only the large scale behaviour of the non-Euclidean space. Moreover, it could be applied to spaces of arbitrary dimensions, a feature that would play a pivotal role in the development of relativity. If we consider the coordinates of two neighbouring points in the Riemannian space under study, the distance between those points depends on functions of the coordinates themselves. These functions, collectively called the metric of the space, can tell everything about the geometry of the space. By means of differential equations involving the metric, the characteristic curves of the space, namely the geodesics, can be obtained. By carrying out somewhat complicated, but straightforward, differential operations on the metric functions, one can determine the intrinsic curvature of the space. In Euclidean space, each of these functions can be reduced to unity, whereas in other geometries they may take on complicated forms. Similarly, the intrinsic curvature of the Euclidean space is zero, whereas it is arbitrary in a non-Euclidean space and can vary from point to point.

The full essence of Riemannian geometry cannot be extracted without the help of tensors, geometric quantities that are generalization of vectors. The basic ideas underlying tensor analysis were present in the works of a gifted high-school teacher Hermann Grassmann. As none paid any attention to his ideas, Grassmann shifted his talents to the study of Sanskrit language and literature, making outstanding contributions to a field totally removed from the sphere of mathematics. In the course of time, tensor analysis grew at the hands of mathematicians like Levi-Civita and Ricci. The Riemannian formalism, buttressed by tensor calculus, brought out the rich texture of geometry and revealed with unprecedented clarity its deep foundations. In parallel with the developments in geometry that found their final fruition in Riemann's work, ideas in regard to the nature of physical space were also undergoing gradual transformation.

Nowadays the common course taken by mathematics, especially within the domain of specialized abstract areas, runs with hardly any contact at all with physical reality. This is totally contrary to the spirit that prevailed throughout the growth and study of Euclidean geometry as well as during the creation of non-Euclidean geometry. As has been noted, for more than two millennia the laws of Euclid were assumed to hold unchallenged sway over the idealized version of physical space. This had to be so, for God had stamped His seal of perfection on the design of the world in the form of the flawless inner structure of Euclidean geometry. Any feeble doubts regarding the applicability of Euclidean geometry to the physical space, that reared their heads from time to time, were put to rest by Immanuel Kant. He declared that the '*a priori* synthetic truths' that exist in the mind independent of experience are those of Euclidean geometry and by necessity the physical world had to follow its rules.

All the same, even before the advent of non-Euclidean geometry and in the face of firmly established traditional view, the Euclidean reign over the physical space was questioned by a handful of daring minds especially during the seventeenth and the eighteenth centuries. George Klügel made the bold assertion that the parallel axiom was but a product of experience and not an *a priori* truth. Ferdinand Schweikart could very well imagine a geometry in which the angles of a triangle did not add up to two right angles. He further conjectured that this 'astral geometry' might be valid on celestial scales. Gauss, whose own work foreshadowed non-Euclidean geometry, tried to test the precise applicability of or deviations from Euclidean geometry by measuring the sum of the angles of the triangle formed by three mountain peaks. The observational error involved was far too large for any conclusion to be drawn from the measurement. Lobatchevsky seriously considered the question whether in fact his newly created geometry governed the physical space. He was able to show that the characteristic length associated with his geometry was million times the size of earth's orbit around the sun. Hence, he concluded that any departure from Euclidean geometry could occur only on astronomical scales. The conclusions of Gauss and Lobatchevsky were thus based on actual numbers and were therefore a step ahead of the imaginative speculations of Schweikart.

The most prophetic speculations in regard to the non-Euclidean nature of the physical space were made by Riemann himself and by the English geometer William Clifford. Riemann believed that the applicability of Euclidean geometry to the physical space was valid only as an approximation; the exact geometry could be ascertained only by astronomical measurements. Further, he wrote: 'Either therefore the reality that which underlies space must form a discrete manifold, or we must seek the ground of its metric relations outside it, in binding forces which act upon it. The answers to these questions can only be got by starting from the conception of phenomena ... This leads us into the domain of another science, that of physics ...'. Clifford made elaborate and bold conjectures which envisaged 'little hills' on the otherwise flat space and visualized the passage of this spatial curvature in the form of waves equivalent in reality to the motion of matter. He concluded 'that in this physical world nothing else takes place but this variation, subject, possibly, to the law of continuity'. The vision of Riemann and Clifford, in which space, non-Euclidean geometry and matter are interwoven to produce the world of experience, would be given concrete shape by the theory of relativity. This brings us to an entirely new chapter that opens on to the present day picture of the universe.

Einstein's world

With the help of his law of gravitation and the three laws of motion, Newton could explain practically everything that was known of the heavens in his time. This extraordinary body of knowledge was presented by Newton in the third book of his *Principia* under the title, 'The System of the World'. Lagrange, a century later, expressed his admiration for Newton in words not untouched with a trace of envy: 'Newton was the greatest genius that ever existed, and the most fortunate, for we cannot more than once find a system of the world to establish.' It was impossible to foresee at the time of Lagrange that a system of the world would be established a second time, with a new theory of gravitation as its base. This new theory, namely Einstein's general theory of relativity, would fuse together space, time and gravitation, and mould the large-scale geometry of the universe.

In trying to elucidate his ideas of motion, both uniform and accelerated, Newton had envisioned absolute space, infinite and all pervading, and absolute time, flowing uniformly without beginning or end. These ideas came under severe attack on both theological and scientific grounds. For instance, Bishop Berkeley condemned 'the dangerous dilemma of thinking either real space is God or else that there is something besides God which is eternal, uncreated, infinite, indivisible, unmutable. Both which may justly be thought pernicious and absurd'. Einstein pointed out that the idea of absolute space and time 'conflicts with one's scientific understanding to conceive of a thing which acts but cannot be acted upon'. The new concepts of Einstein's relativity would remedy these and other shortcomings of the Newtonian ideas of space and time.

The special theory of relativity showed that both spatial and temporal measurements varied among observers moving with uniform velocities with respect to one another. A coherent scheme incorporating those relative variations could emerge only if time were combined with space on an equal footing. As the mathematician Hermann Minkowski, the one time teacher of Einstein at the Zurich polytechnic, prophesied in 1908, 'from now on space by itself and time by itself are destined to sink completely into shadows, and only a kind of union of both will retain an independent existence.' Spatial points are now replaced by 'events' in spacetime, each event characterized by three spatial coordinates, indicating *where* the event took place, and one time coordinate specifying *when* it occurred. In our good old three-dimensional space the distance between two points could be computed by taking the differences in the corresponding coordinate values, then squaring and adding them, provided we used the usual mutually orthogonal Cartesian coordinates. One can cover a flat piece of paper, for instance, with such coordinates, but not a curved surface like a sphere. The three-dimensional physical space was assumed to admit such coordinates which in esssence meant that the space was Euclidean. Now in the four-dimensional spacetime, *when there is no gravitation*, one can again define a similar coordinate system covering all of spacetime. However, the four-dimensional separation between two events is obtained by *adding* the squares of the spatial coordinate differences but *subtracting* the square of the time coordinate difference. This gives the *invariant* four-dimensional separation; that is, though the coordinate values and their differences may vary among different coordinate systems representing different observers, the above quantity retains the same value. Notice that this invariant length, or line element as it is called, has a form similar to that on a flat surface. Furthermore, the paths of free particles in four dimensions are straightlines, since velocity is constant in the absence of force according to Newton's first law of motion. As it turns out, these free particle trajectories are the geodesics of the four-dimensional spacetime we are considering. Again, this is perfectly analogous to the flat sheet of paper on which geodesics, curves giving the shortest distance between two points, are straightlines. The conclusion we have arrived at is this: the physical spacetime of special relativity, which *does not include gravitation*, has a flat geometry.

How does one introduce gravitation into this framework and with what consequences? One can only marvel at Einstein's genius for accomplishing this task taking as his starting point one of the simplest observations that had been known for centuries. Galileo is supposed to have demonstrated that all bodies fall to the ground with the same acceleration irrespective of their masses by dropping two dissimilar objects from the tower of Pisa. Because of this fundamental property of gravitation, observed Einstein, one could enunciate an 'equivalence principle' that gravity at a point could be replaced by an accelerated frame. No physical phenomenon would find any difference between the two situations. Far away from the earth, where gravity is almost zero, propel a rocket 'upwards' with the same acceleration as that due to gravity on the earth. Then all objects, when dropped inside the rocket, seem to fall with this acceleration as the floor of the rocket rushes up to them. This way gravity has been simulated by an accelerated frame. On the other hand, when there is already gravity present, it can be annulled by acceleration. This

Fig. 10.7. Gravity is simulated within an accelerated rocket. Gravity, already existing, is annulled inside a freely falling frame like the Einstein elevator.

is the famous thought experiment involving the Einstein elevator. Cut the cables of the elevator and let it fall freely under gravity. Inside the elevator, objects seem to float as if there were no gravity at all, because everything including the observer is falling freely at the same rate. What has all this got to do with spacetime? Everything. Remember, when gravitation was absent spacetime was flat. The element of spacetime inside the freely falling elevator is flat since gravity has been eliminated. Suppose we have a gravitational field which varies all over space and possibly in time too. Then at each point of spacetime one can think of a freely falling elevator in which spacetime is a small flat patch. But, because of the varying nature of the gravitational field, these elevators will be falling with different accelerations in different directions. The consequence of this is that the corresponding elementary flat spacetime patches, when meshed together, will in general form a globally curved spacetime which incorporates the non-constant gravitational field. Or in short, the presence of gravitation engenders a curved spacetime. One can very well have a gravitational landscape made of 'small portions of space ... of a nature analogous to little hills on a surface which is on the average flat', as Clifford had envisioned!

Fig. 10.8. In an arbitrary gravitational field the acceleration of freely falling frames varies from point to point both in magnitude and direction. Because of free fall, gravity is eliminated and the local element of spacetime in each frame is flat. These flat patches of spacetime mesh together to form in general a global curved spacetime in a manner similar to the two-dimensional analogue.

In the absence of gravitation, particles and light rays travel along straight lines which are geodesics of a flat spacetime. When gravitation is present, the paths followed by them are nothing but the generalised 'straight lines', the geodesics, the natural curves adapted to the curved spacetime. There is no longer any need to think of a Newtonian force deviating the particles from straight paths. The particles have no choice but to follow the curves chalked out for them by the curved space. But the fabulous surprise here is that this must be true not only of material particles but also of light rays. Light rays must bend in a gravitational field, a phenomenon unheard of in Newtonian physics. This effect, first verified by Eddington and firmly established by subsequent astronomical observations, is an extremely important one on cosmic scales. After all, observations related to the universe are carried out by means of light rays. The dominant gravitation associated with the large-scale structure of the universe leaves its signature on light, thereby modifying observations. These observations, when suitably interpreted, can reveal a great deal about the universe that light has traversed.

What is the final source of the curvature of spacetime? Obviously, it is the same source that produces gravitation, namely matter or energy of any kind. This could be ordinary matter, radiation, energy associated with rotation,

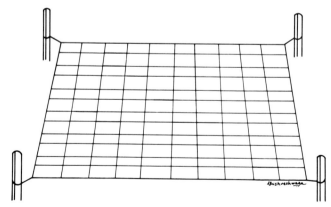

Fig. 10.9. Acrobats' net as a two-dimensional analogue of spacetime. In the absence of gravitating matter, the surface of the net is flat. The ropes form an orthogonal Cartesian coordinate grid.

magnetic fields and so on. To visualize how mass affects spacetime, imagine a tightly stretched net like the one acrobats use. The two-dimensional surface defined by the net is flat. The mutually orthogonal ropes of the net form a Cartesian coordinate grid. The distance between adjacent points is given by the sum of the squares of the coordinate differences. When gravitating mass, in the form of a fallen acrobat, appears on the net, the latter along with the ropes becomes 'curved'. The presence of matter has now curved the spacetime. Points on the net can still be marked by the intersections of the ropes as before. But the coordinate grid of the ropes is now curved and the distance between neighbouring points is no longer given by the simple 'sum of the squares' formula. Coordinate functions have to be fed into the

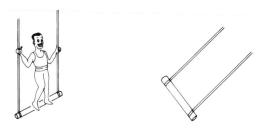

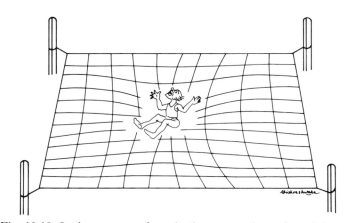

Fig. 10.10. In the presence of gravitating matter the surface of the net, analogous to spacetime, becomes curved. The ropes no longer form a global Cartesian grid. The net or the spacetime has now a general Riemannian geometry.

generalization of the formula which was valid in flat space. In other words we now have a general Riemannian geometry describing the curved spacetime. In four-dimensional spacetime, one needs in general ten such metric functions. These are obtained by solving the Einstein field equations. They relate the geometry of the spacetime to the energy and momentum distribution of the sources producing gravitation. These equations are ten in number and quite complicated in structure. Their solutions, once obtained, provide all the relevant information on the geometry of the curved spacetime and therefore on the gravitational field the spacetime describes.

Let us continue with the analogy of the net a little further. Where the source of gravitation, namely the acrobat, is located, there is a depression in the surface defined by the net. If an object is let go in this region it will roll down towards the acrobat because of the depression; the exact path depends on the initial velocity imparted to the object. In the four-dimensional spacetime this is equivalent to the motion of particles along geodesics adapted to the curved geometry produced by the gravitating source. The net becomes less and less curved, flattening out gradually, as one moves outwards from the depression. Likewise, the spacetime curvature dies down continuously as the distance from a localised matter distribution increases; far away from the source of gravitation the spacetime is again almost flat. Any motion of the acrobat is transmitted to the net and, in particular, if he moves up and down rhythmically, waves are set up in the net. Analogously, if any material object executes periodic motion, the curvature of spacetime can propagate as a wave. This is the phenomenon of gravitational waves, a purely general relativistic effect. If the mass on the net is enhanced without spreading it out, the sagging of the net can increase sharply. One can imagine a situation when objects near enough to this depression can roll down but will be unable to climb back. Such a phenomenon is indeed possible in the gravitation dominated spacetime of a black hole.

The black hole is again a purely general relativistic entity. It is an extreme and a fascinating example of what curved geometry of the physical spacetime can do. Things, including light, can enter a black hole, but cannot come out of it! How can we understand this? In the absence of gravitation, or as on earth when its effect is not very high, we are familiar with the fact that a wavefront of light can cross us, but we cannot recross it in the other direction. To do this we will have to chase it at a speed greater than that of light which is impossible. In other words, one can cross a wavefront in only one direction. If, then, a closed wavefront – or a geometric surface with its properties – can be created frozen in space, then one can fall through it but cannot come back recrossing it in the other direction. This is precisely the amazing feat performed by gravitation. The spacetime curvature is such that a closed geometric surface having no physical content but with the properties of a wavefront is frozen in space acting as a one-way membrane. And this is the black hole.

Of course the grandest application of spacetime geometry lies in the description of the universe as a whole. The material building blocks of the universe are taken as the galaxies or the clusters of galaxies. The observational input that goes into the construction of cosmological models is that the universe, on the scale of clusters of galaxies, appears the same in all directions, that is the universe is isotropic as we see it. This does not tell us how the universe looks like at and from other points. However, if one invokes the so-called Copernican principle that we occupy no special place in the universe, then this principle combined with the observed isotropy tells us that the universe is homogeneous. This means that on a large scale the physical properties like the density and the distribution of galaxies are the same at all points of the universe. This elementary requirement, surprisingly enough, sharply narrows down the possible geometries of the cosmic spacetime. The models that emerge thus are of only three types. The intrinsic geometry of the spatial section at any given moment of time varies among these models. One of them – the closed model – has closed spatial section, the spatial geometry curling up on itself. One can think of the surface of a sphere as a two-dimensional analogue, but it is impossible to visualize a closed three-dimensional space. The other two are open models infinite in their spatial extensions. The two-dimensional analogues of these spatial sections are the flat plane and the hyperboloid, the latter resembling the surface of a saddle. Which of the three models corresponds to the actual universe depends upon the amount of matter content of the latter. If the density of all types of matter and energy put together exceeds a critical value, the spacetime curvature can become high enough to produce a closed universe. If the density is equal to or less than the critical value, the spatial sections will be flat or hyperbolic respectively. All the three models start from an initial singular state in which

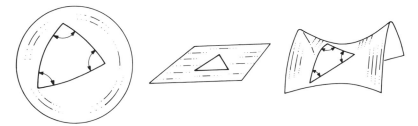

Fig. 10.11. The two-dimensional analogues of the spatial sections of the three homogeneous cosmological models: the sphere, the plane and the hyperboloid. The sum of the angles of a geodesic triangle is respectively greater than, equal to and less than two right angles in these three cases.

density, temperature and curvature are all infinite. The universe expands after this 'big bang', the galaxies flying away from one another. The spatially closed universe can expand only up to a maximum size and then it contracts to another singular state or the 'big crunch' to resume its expansion once again. It can thus go through cycles of expansion and contraction. The other two open models, on the other hand, expand indefinitely since the matter content, and consequently the gravitational attraction, is not sufficient to slow them down. The mathematical structure of the geometry and the evolution of these three cosmological models are both simple and beautiful. Their geometries have provided a firm framework for describing various physical processes occurring on cosmological scales. The theoretical studies of these processes, in turn, have been spectacularly successful in explaining the observed properties of the universe at large.

However, which of the three types of cosmological models actually corresponds to the universe we live in is an open question. The answer to this question will not only specify the geometry of our universe, but will forecast its fate as well. This issue can be settled only by refined observations. Einstein's theory only offers the possible geometrical models of the universe, but does not decree a unique choice. This ultimate prerogative belongs to physical reality as incorporated in the properties of the constituents of the universe.

Because of Einstein's vision, geometry has found its true role in the description of nature. Space is no longer a passive arena in which gravitational phenomena unfold to the rhythm kept by time. Space and time combine to form the fabric of the world whose curvature is identified with gravitation. The structural bases of the universe built up almost entirely on conceptual grounds by general relativity have to a considerable extent revived the original world view of the Greeks: geometry has been recast as the cosmic framework and the supremacy of the intellect in arriving at the truths underlying physical reality has been restored. This led Einstein to remark: 'It is my conviction that pure mathematical construction enables us to discover the concepts and the laws connecting them, which gives us the key to the understanding of Nature ... In a certain sense, therefore, I hold it true that pure thought can grasp reality as the ancients dreamed.'

The far horizons

Whereas Einstein's general relativity fulfilled to a large measure the ancient Grecian dream, his own dream of uniting the fundamental forces of nature has remained far from realisation. His strenuous efforts towards building a unified theory extended over a period of more than thirty years beginning in the nineteen twenties and continuing till his last days. Einstein's aim was to blend together gravitation and electromagnetism within a geometric framework akin to that of general relativity. His hope was that a fully integrated theory might even embrace quantum mechanics remedying its statistical character which was repugnant to his philosophy of nature. Einstein's indomitable spirit and his relentless labours bore no fruit in the end. All the same, they have become a source of inspiration and a shining example in the continuing quest for the sacred monolith. The central theme of unification as envisaged at present is, however, quite different from Einstein's original idea. In his days the physicists had just glimpsed the inner workings of the weak and the strong forces responsible respectively for radioactive decay and nuclear binding. Consequently, these two forces did not enter into Einstein's vision of unification at all. Today the cherished goal is to combine *all* the four basic forces of nature – gravitation, electromagnetism, the weak and the strong interactions.

As things stand at present, the unification of three of the four forces may be considered to have been accomplished already. This is achieved through the 'standard model' which consists of the electro-weak force – a coherent combination of electromagnetism and weak interaction – coexisting on equal footing with strong interaction for which a viable theory is available. Furthermore, theories have been built, but not yet experimentally tested, that integrate all the three forces into a single scheme, the so-called Grand Unified Theories. Perhaps, the resounding epithet should have been reserved for a theory that truly unified *all* the four forces including gravitation, and not just three. Gravitation, the force that is the weakest in nature but the toughest to tackle, has resisted all attempts to bring it in line with the other interactions. Nevertheless, intense efforts aimed at achieving the ultimate goal continue demanding an unlimited investment of energy, ingenuity and, above all, hope. Almost all the theories proposed introduce at some point or the other a geometric picture similar to that of general relativity. For instance, some of these theories are patterned after the one proposed by Theodor Kaluza and developed by Oskar Klein during the twenties. They had hit upon the rather unusual idea of associating electromagnetism with a fifth dimension that had the form of an extremely small circle. Even Einstein was impressed with this novel construction, even though he dismissed the notion of invoking more than the conventional four dimensions in his own unification programme. However, the modern Kaluza–Klein type theories build up spaces of as many as eleven dimensions in order to fit in the weak and the strong interactions as well. While four of them constitute the usual spacetime, the additional dimensions at each point are unimaginably minute, curled up like circles and inaccessible to direct observation. These and other theories suffer from some incurable malady or the other. In their present forms, they are a far cry from a satisfactory union of the four forces. How far they will succeed in approaching the final goal, when further refined and developed, none can tell at the present stage. One cannot even rule out the sad possibility that they may prove to be nothing more than sand castles built on the seashore only to be washed away by the shifting tide. Some vestiges of their basic concepts may survive in the final unified theory which could very well turn out to be essentially geometric in character.

One of the most important features shared in common by the three forces that have been successfully integrated is that each of them can be described separately by a relativistic quantum field theory. Such a theory is the result of formulating the quantum description of the force field subject to the requirements of the special theory of relativity. The flat spacetime of special relativity is the arena in which the quantized fields perform and co-mingle. But, what about gravitation? Once again, quantization of gravity has remained an elusive, ever-receding goal. The obvious approach to attain this end would be to quantize general relativity. But, here one is faced with a peculiar but fundamental difficulty. Unlike as in the case of the other forces, there is no longer a passive arena of action at one's disposal. The curved spacetime of general relativity by itself represents gravitation – the very field to be quantized. Because of this deep-rooted conceptual barrier, as well as other technical difficulties, all attempts made time and again to quantize general relativity, backed by imagination and skill, have met with futility and frustration. This of course does not deter theorists from building partially complete models of quantum gravity and speculating on the possible consequences of a full theory. Quantum effects of gravitation would show up at extraordinary small scales of spatial dimensions, roughly about twenty powers of ten smaller than the nuclear size. At such small dimensions spacetime is conjectured to lose its normal smooth geometry and take on a structure resembling the vibrant foam on a turbulent sea. How far such conjectures will be borne out in the end, only time can decide.

In the context of his own philosophy of unification, Einstein wrote in his *Autobiographical Notes*: 'I have learned something else from the theory of gravitation: No ever so inclusive collection of empirical facts can ever lead to the setting up of such complicated equations. A theory can be tested by experience, but there is no way from experience to the setting up of theory.' Both quantum gravity and the unified field theory will no doubt be built on purely conceptual grounds. But where can one find the 'experience' associated with these two theories? As has been already mentioned, quantum gravity becomes important at distance scales of about hundred-billion-billionth the nuclear size. In the micro-world we are dealing with, each distance scale is associated with an energy that is required to probe structures of that size. At the atomic distance scale this energy corresponds to a few electron volts. As the distance reduces to the infinitesimal level of quantum gravity, the associated energy rises to the awesome value of some billion-billion-billion times that corresponding to the atomic dimensions. It is also at such energies that the four forces are expected to merge totally and become one. As the energy decreases, the seed sprouts and the forces branch out one by one. The only way to 'experience' the influence of quantum gravity and of the confluence of forces is to travel way back to the beginning of the universe.

General relativity, while successfully elucidating the geometric structure and the evolution of the universe, has at the same time brought with it the problem of the cosmic origin. The theory predicts the inevitable existence of a primordial singularity where not only physical quantities like density and temperature blow up, but also the curvature of spacetime geometry itself becomes infinite. This is where intuition, physics and mathematics – all three alike – break down. It is the ardent expectation of the theorists that quantum effects of gravitation would prevent the actual occurrence of such a singular state. The nature of these quantum effects lies in an area of total darkness in the absence of an exact theory of quantum gravity. However, when the entire universe is in a shrunken state confined to the inconceivably small distance scales of quantum gravity, its temperature and the corresponding energy would be unimaginably high. This is precisely the domain in which the fusion of all four forces assumes a paramount role. In all probability, the two aspects of gravity – its quantum character and its merging with other forces – would be inseparably intertwined to produce their cumulative effects. At this stage one will have reached out to the very birth of the universe. This is where questions belonging to the sphere of mysticism migrate into the realm of physics. What is the nature of space and time near the event of cosmic birth? Can one even continue to entertain these conventional concepts any longer? Does the birth of the universe imply the beginning of spacetime? If so, how can one think of the phase before this beginning if there was neither space nor time? Was the universe created from nothing? What is nothing? Only when the twin towers of quantum gravity and the ultimate unified theory have been built to the full, possibly on a common foundation, can one hope to seek answers to these questions concerning the meaning – and perhaps the very genesis – of spacetime at the point of creation.

Centuries of observation and thought have unravelled the patterns of cosmic design, sometimes in slow stages and sometimes in brilliant bursts. More than two thousand years stretch between Euclid and Einstein, between the early glimmer of a geometric structure for the universe and its final construction. But deep questions concerning the birth of the universe itself remain to be answered. It is a wonder that such questions can even be posed within the framework of physics and mathematics. The answers to these questions may call for radically new concepts in physics and entirely new types of mathematics, both of which may lie beyond our imagination at present. How long we will have to wait for these answers, no one can predict. Hopefully we will come to know them in our own lifetime. For all we know, somewhere there is already a child of twelve deeply absorbed in his little holy book on non-Euclidean geometry!

References

Bell, E.T. 1965. *Men of Mathematics*, Simon and Schuster.
Cornford, Francis MacDonald 1973. *The Republic of Plato*, Oxford University Press.
Ghyka, Matila 1985. *The Geometry of Art and Life*, Dover.
Greenberg, Marvin Jay 1980. *Euclidean and Non-Euclidean Geometries*, Freeman.
Hamilton, Edith and Huntington, Cairns (ed.) 1961. *Plato – the Collected Dialogues*, Pantheon Books.
Hildebrandt, Stefan and Tromba, Anthony 1985. *Mathematics and Optimal Form*, Scientific American Library.
Kline, Morris 1972. *Mathematical Thought from Ancient to Modern Times*, Oxford University Press.
Koestler, Arthur 1960. *The Watershed*, Doubleday.
Kramrisch, Stella 1946. *The Hindu Temple*, University of Calcutta.
Thibaut, G. 1875. *On the Śulvasūtras, J. Asiatic Soc. Bengal* pp. 227–75 (Reprinted in *Studies in the History of Science in India* Vol.II, ed. D. Chattopadhyaya, Editorial Enterprises, New Delhi 1982).
Thompson, D'arcy Wentworth 1985. *On Growth and Form*, Cambridge University Press.

11

The origin and evolution of life

CYRIL PONNAMPERUMA

The question

'How did life begin?' is a question that has been uppermost in the mind of man from almost the very dawn of time. 'Even the formulation of this problem is perhaps beyond the reach of any one scientist. For such a scientist would have to be, at the same time, a competent mathematician, a physicist, and an experienced organic chemist. The same person should have a very extensive knowledge of geology, geophysics and geochemistry, and besides all this, be absolutely at home in all biological disciplines. Sooner or later, this task would have to be given to groups representing all these faculties and working closely together, theoretically as well as experimentally.' Such was the view professed by J.D. Bernal in 1949. However, today we may have a reason to be more optimistic. For the first time in human history, the sciences which arose as separate disciplines are seen fused together, and our own generation has witnessed the work of the interdisciplinary sciences of biophysics, biochemistry, molecular biology, and astrochemistry.

The scientific basis

Three factors have made the scientific approach to the question, 'How did life begin?' possible, not only theoretically, but also experimentally: the astronomical discoveries of this century, recent biochemical advances, and the triumphs of Darwinian evolution.

The astronomical discoveries have relegated our earth to the corner of a universe made up of billions of stars. The study of the heavens by present-day telescopes has revealed the presence of more than 10^{23} stars. Like our own sun, each one of these stars can provide the photochemical basis for plant and animal life. Two factors become abundantly clear: there is nothing unique about our sun, which is the mainstay of life upon this planet and there are more than 10^{23} possibilities for life in the universe. In the light of numerous possible restrictive conditions, conservative estimates made by astronomers put the number of possible sites in the universe for life at about 1% of these possibilities. More optimistic calculations suggest 50%. In the light of such calculations, the earth then appears to be just one place in the universe where a successful experiment has been performed. If the laws of physics and chemistry are universal laws, one may then extrapolate from earth to elsewhere in the universe.

When we turn from the stars to the earth, we find that recent biochemical discoveries have underlined the remarkable unity of living matter. The chemical composition of living matter is qualitatively different from that of the physical environment in which they live. Most of the chemical components of living organisms are organic compounds of carbon, in which the carbon is relatively reduced or hydrogenated, many also contain nitrogen. The organic compounds present in living matter occur in extraordinary variety and most are extremely complex being macromolecules with very large molecular weights. These include the proteins and nucleic acids.

However, the immense diversity of organic molecules in living organisms is ultimately reducible to a surprising simplicity since the macromolecules are composed of simple, small building block molecules strung together in long chains. The different types of proteins consist of long covalently linked chains of amino acids, small organic compounds of known structure. Only twenty different amino acids are found in proteins, but they are arranged in many different sequences to form proteins of many different kinds. Similarly, the nucleic acids of which there are two kinds, DNA and RNA, are constructed from a total of eight different building blocks, the nucleotides, four of which are building blocks for DNA and four for RNA. Moreover, the twenty different amino acid building blocks of proteins and the eight different nucleotides of nucleic acids are identical in all living species, leading to the inescapable conclusion that all living organisms must have had a common chemical ancestry.

Another factor on which the scientific study of the origin of life can be based is the triumph of the Darwinian theory of evolution. According to Darwin, the higher forms of life evolved from the lower ones over a very extended period in the life of this planet. Fossil analysis has shown that the oldest known forms of living systems may be about three and one half billion years old. From geochemical data, we know that the earth is 4.6 billion years old. Life appeared on this earth sometime between the 4.6 billion years of age of the earth and the 3.5 billion years where we have convincing evidence for the presence of a microbiota.

Chemical evolution

Chemical evolution may be considered to have taken place in stages, from inorganic to organic and from organic to biological. The first stage of chemical evolution perhaps began with the very origin of matter. In a series of cataclysmic reactions during the birth of a star, the elements of the periodic table must have been formed. Studies of nuclear synthesis have shown us that the origin of the elements can be traced from the 'big bang' when our universe was first formed through the birth of stars, to the first generation stars where hydrogen burning would occur, finally giving rise to elements up to iron through the process of helium, carbon, and silicon burning. Then would follow the second generation stars as a result of the supernovae which give rise to the heavier elements beyond iron in the periodic table. When the solar system was being formed, presumably from material available through an explosion of the supernova – our sun is indeed a second generation star – the very elements necessary for life were available in the primordial nebula.

Within this framework, then, life may be considered to be a special property of matter, a property which arose at a particular period in the existence of our planet, and which resulted from its orderly development.

Scientific thinking

To the Russian biochemist, A.I. Oparin, more than to anyone else today, we owe our present ideas on the scientific approach to the question of the origin of life. In clear and scientifically defensible terms, he pointed out that there was no fundamental difference between living organisms and brute matter. The complex combination of manifestations and properties so characteristic of life must have arisen in the process of the evolution of matter. In 1928, the British biologist Haldane expressed his own ideas on the origin of life. He attributed the synthesis of organic compounds to the action of ultraviolet light on the earth's primitive atmosphere. He suggested that the organic compounds accumulated till the primitive oceans had the consistency of a primordial soup. Twenty years after the appearance of Haldane's paper, in the *Rationalist Annual*, Bernal theorized before the British Physical Society in a lecture entitled, *The Physical Basis of Life*. 'Condensations and dehydrogenations are bound to lead to increasingly unsaturated substances, and ultimately to simple and possible even to condensed ring structures, almost certainly containing nitrogen, such as the pyrimidines and purines. (Pyrimidines and purines are components of nucleotides.) The appearance of such molecules made possible still further synthesis.' The primary difficulty, however, of imagining processes going this far was the extreme dilution of the system, if it is to take place in the free ocean. The concentration of products is an absolute necessity for any further evolution.

The raw material

The raw material from which the building blocks of life have evolved consists of the chemical elements of the periodic table. Examination of the crust of the earth, the oceans, and the atmosphere provide us with information about the abundance of these elements on the earth. Data on the elemental composition of matter beyond the earth comes from several sources. The spectroscopic analysis of light from stars reveals the nature of the elements in them. The developments of the science of radio astronomy has provided us with the microwave technique of detecting various elements and excited species in intergalactic space. Cosmic ray particles can supply us with samples of extraterrestrial matter. Meteorites and lunar samples have given us valuable knowledge of the composition of our solar system. A reasonable and consistent picture of the abundance of the elements in the universe can thus be obtained.

The nature of the primitive atmosphere

A true understanding of the nature of the earth's primitive atmosphere is a logical starting point for any discussion of the problem of the chemical origin of life. While it is difficult to answer by direct observation the questions that arise, evidence available from a number of sources indicate that a reducing atmosphere gradually gave way to the oxidized atmosphere of today. Information gleaned from astronomy, astrophysics, chemistry, geology, meteoritic studies, and biochemistry can be used to elucidate this problem. The present rarity of the noble gases in the earth's atmosphere in comparison to their distribution in the universe indicates that the primitive atmosphere of the earth was lost, and that the atmosphere that is generally described as primitive was secondary in origin. This atmosphere must have resulted from the outgassing of the interior of the earth during planetary accretion. The secondary atmosphere must have been very similar to the first, and it is this atmosphere which would be considered in our discussion of the primordial atmosphere of the earth.

Energy sources

The energies available for the synthesis of organic compounds under primitive earth conditions were ultraviolet light from the sun, electrical discharges, ionizing radiation, and heat. It is evident that sunlight was the principle source of energy. Solar radiation is emitted in all regions of the electromagnetic spectrum. A great deal of this radiation, especially in the ultraviolet region, is shielded today by the ozone layer in the upper regions of the atmosphere. Since reaching its main sequence, the sun has been very stable. The temperature of the sun's surface four and one half billion years ago was almost as great as it is today. With this information, the solar flux at different wavelengths has been calculated. These figures clearly demonstrate the dominant role that ultraviolet light must have played among primitive energy sources. The use of ultraviolet light should be considered as far more important than any other sources of energy that have been used in simulation experiments in the laboratories. However, on account of the intrinsic difficulties that have been involved in using short wavelength ultraviolet light, not many experiments have been performed with this energy source.

Next in importance as a source of energy are electrical discharges such as lightning and corona discharges from pointed objects. These occur close to the earth's surface, and therefore could have effectively transferred the reaction products to the primitive oceans. Since lightning can be easily simulated in the laboratory, many experiments have been performed using this form of energy.

The principle radioactive sources of ionizing radiation on the earth are potassium-40, uranium-238, uranium-235 and thorium-232. Potassium-40 seems to be quantitatively more important than the other three sources at present.

Heat from volcanoes was another form of energy which may have been effective. It is reasonable to expect that volcanic activity was more prevalent on the primitive earth than today. Besides, heat would also have been available from hot springs around boiling mud pots.

Chemosynthesis by meteorite impact on planetary atmospheres has been suggested as a possible pathway for primordial synthesis of organic matter. The reaction is probably a result of the intense heat generated momentarily in the wake of the shock wave following the impact.

Experimental work in the laboratory

The student of chemical evolution has tried in the laboratory to recreate Darwin's 'warm little pond'†. That the gaseous components now known to be present in the primitive atmosphere can be the precursors of

---

† Darwin, in a letter to his friend Hooker, wrote 'in some warm little pond, with all sorts of ammonia and phosphoric salts, light, heat, electricity, etc. present ... a protein compound was chemically formed ready to undergo still more changes'.

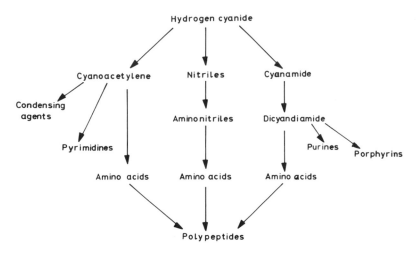

Fig. 11.1. Chemical evolution of biomolecules from hydrogen cyanide under simulated primitive-earth conditions.

organic compounds is today well supported by laboratory studies. Among the early experiments on the abiotic origin of organic molecules were those carried out in 1953 by Stanley Miller. Miller exposed a mixture of methane, ammonia, water, and hydrogen to an electric discharge from tesla coils for about a week. Several organic compounds were formed. Among these were the amino acids, glycine, alanine, aspartic acid, and glutamic acid.

Many different forms of energy or radiation lead to organic compounds from simple gas mixtures. Several hundred different organic compounds have been formed in such experiments carried out by Ponnamperuma and his associates and also by other workers. These compounds include representatives of all the important types of molecules found in cells as well as many not found in cells. All the common amino acids found in proteins, the nitrogenous bases adenine, guanine, cytosine, uracil and thymine, which serve as the building blocks of nucleic acids, and many biologically occurring organic acids and sugars have been detected among the products of such primitive-earth-simulation experiments. It appears quite likely that the primitive ocean was indeed rich in dissolved organic compounds, which may have included many or all of the basic building-block molecules we recognize in living cells today.

Synthesis of large molecules

In putting two amino acids together to form a dipeptide or two nucleotides together to form a dinucleotide, a dehydration-condensation is necessary. Dehydration could have been achieved by the result of the action of heat on the primordial solution of organic molecules by the evaporation of such a mixture by thermal action. In his lecture *The Physical Basis of Life*, Bernal had suggested that the organic components from the ocean could have been brought from the ocean to the shoreline, absorbed on clay and condensation could have taken place. The Bernalian idea has been tested in the laboratory with encouraging results.

Yet another possibility would be to see whether the reaction could take place in water. Since the earth is a very wet planet and if it can be demonstrated that these reactions could take place in water, then a greater opportunity for the formation of polymers in the prebiotic conditions would exist. In living processes, such a system does indeed take place. Enzymes are used to overcome the energy barriers. In the prebiotic system, what could be necessary would be predecessors of such enzymes or primitive catalysts. A number of experiments have been performed in which both hypotheses have been tested with very satisfactory results. The dehydrating condensation of amino acids to yield polypeptides can be brought about by heat, by organic condensing agents, by polyphosphates and phosphoramidates or by clay and other mineral catalysts. Primitive polypeptides called proteinoids form readily when amino acids mixtures are heated above 100 °C. Such proteinoids have a nonrandom amino acid composition, show a number of properties of modern proteins and also have rudimentary catalytic activity. Nucleotides and polynucleotides also form under mild primitive earth conditions, particularly in the presence of polyphosphates or phosphoramidates. Simple polynucleotides can serve as templates for the abiotic formation of complementary polynucleotides.

Interaction between nucleic acids and proteins

Primitive cell-like structures may have formed by the process of coacervation or by micelle formation driven by hydrophobic interactions, the tendency of the surrounding water molecules to seek the position of maximum entropy. A.I. Oparin has postulated the formation of coacervate droplets composed of polymers, into which primitive catalysts may have been incorporated. More recently S.W. Fox has described proteinoid microspheres, which exhibit many aspects of cell like behaviour. Such structures were visualized as developing in the absence of nucleic acids which may have been acquired later.

The other general hypothesis is that a primitive gene was required before proteins were added, an idea supported by the structure of modern viruses, the general importance of nucleotides in modern biochemistry and the capacity for self-replication. In either case, some means for recording or coding and thus, self-replication was required, as well as the capacity for further evolution. Development of the genetic code may have been the basic event in the origin of life.

Many studies have shown that a variety of interactions can occur between nucleic acids and amino acids depending upon composition, conformation, state of polymerization, and environment of the reacting species. These studies have led to the conclusion that a degree of specificality does exist, although its origin has not yet been elucidated. When the two reacting species are simple, one cannot expect to observe specificity of the sort implied in the biological use of the term. What one can look for at the simplest level is evidence for selectivity of some sort, cases where the strength of binding of an amino acid to a nucleotide under given conditions is to some extent a function of the composition of both interacting species. While results of this sort from simple protein-nucleic acid systems would not be spectacular from a biological point of view, they are nevertheless a first step in any systematic study of a role for amino acid/nucleic acid interactions in the evolution of living systems. Working with monomeric species in aqueous media can permit the effects of individual factors to be assessed, and to provide basic information necessary to interpret the more complicated polynucleotide amino acid interaction. If interactions exist, what are they? Are they selective? Can they lead to nucleic acid directed protein synthesis, or help one to understand the system? Can they be applied to understand a more complex system? Can they lead to the addition of constraints which can achieve genetic coding in a defined prebiotic system?

Another possible system would be to look at the biological process as we see it today in living organisms. This could then be looked at in its more simple form and examined in such a way as to see whether the most primitive of the nucleic acid components in the present translation apparatus could have been effective in a much more simply organized translation and replication process.

The analytical approach

The laboratory studies have established that the molecules necessary for life can be made in the laboratory. A question of paramount importance is whether this process did indeed take place on the earth or elsewhere in the universe. Can we go back in time to the earliest stages of the earth, of the solar system and dig up some evidence for the presence of prebiologically synthesized organic molecules? Many attempts have been made to resolve this problem. When did the transition occur from the chemical to the biological system? As the biological record is pushed further back, the period in which chemical evolution could have taken place is compressed. At one time it was thought that no life existed before the Cambrian period 600 million years ago. Paleontologists had argued that there were no skeletons, therefore there was no life. However, the work of the micropaleontologists has pushed the presence of life much further back than 600 million years ago. At one billion years ago occurs the Bitter Springs formation in Australia, in which microstructures have been observed. From three to three and half billion is the Swaziland sequence in South Africa where bacteria-like microstructures have been identified. These microfossils would be the remains of the oldest living entities identified upon the earth.

Molecular fossils

Organic geochemical analysis of Precambrian sediments attempts to determine the nature of the depositional paleoecosystem through the characterization of molecular carbon constituents. The primary goal in most of these studies is to isolate and identify biological markers, chemical fossils which by their molecular configuration and isotopic abundances indicate biological activity. Specific chemical fossils may be characteristic of specific metabolic pathways. The origin of life, the development of bacterial photosynthetic pathways, the appearance of eukaryotic organization, and other events of biochemical evolution could be preserved in the geologic record, not only as microfossils, but also as chemical fossils of discrete origin. However, two problems occur in translating Precambrian chemical fossils into a representation of a paleoecosystem: the alteration of primary organic compounds and postdepositional contamination. In recent sediments, the extractable components are similar to those biochemically synthesized by contemporary microorganisms. The primary organic compounds, however, are rapidly altered by bacterial degradation and the digenetic process. Biopolymers are rarely hydrolyzed to their components. Functional groups are easily replaced by hydrogen, and molecular configurations are destroyed through rearrangement and racemization. With increased temperature, thermal cracking and condensation of organic compounds, kerogen, an insoluble micromolecule with no defined structure, may be formed. A vast portion of organic carbon in the crust of the earth existed in soluble kerogen. Highly metamorphic temperatures altered the kerogen further, until only carbon–carbon bonds are present in the form of graphite. The molecular information is completely lost, and isotopic information may be obscured as well. Organic compounds may have been introduced into Precambrian sediments after deposition. Seemingly impermeable chert is sufficiently porous and permeable to groundwater to contribute significant amounts of recent organic contaminants. On the other hand, the insoluble carbonaceous material which may well constitute 98% or more of the total carbon in a

Precambrian sediment, is generally considered to be syngenetic, as it is physically too large to be mobile.

Certainly more information can be derived from the soluble fraction than from the kerogen or graphite. Yet it may be virtually impossible to distinguish whether the observed molecules date from the time of deposition or were introduced earlier. In exploring the possibility of using molecular fossils as criteria for the presence of life, therefore, one has to be extremely cautious about contamination and diagenesis.

In the work on the Greenland sediments dated as 3.8 billion years old, no attention has been paid to the extractable organics. However, the presence of graphite has been carefully studied and most of the graphite samples seem to give us an isotope fractionation which falls in the biological range. However, there are some samples which may give anomalous figures. In the absence of further evidence to the contrary, these figures may suggest that life on earth is as old as the oldest rocks. If this is indeed the case, it may not be possible to find on earth, evidence of any of the prebiological processes that we have attempted to verify in the laboratory. Evidence for such a process may have to be looked for in other places in the solar system – the moon, the meteorites, and the surface of planets.

Analysis of lunar samples

The Apollo mission provided a unique opportunity for the analysis of samples brought back from the moon. Samples from the Apollo mission 11–16 were analysed in the laboratory. These studies included an examination of the lunar material for total carbon, organic carbon, isotope fraction, microfossils, and mineralogy. Sequential treatment of this sample by benzene, methanol, water, and hydrochloric acid provided extracts for examination by chromatographic and spectromatic methods. To minimize contamination, the analyses were carried out in a clean laboratory with filtered air, and the entire sequence of solvent extractions of the lunar dust was accomplished in a single glass vessel. The total carbon determined by measuring the volume of CO_2 evolved when a one gram sample was outgassed at 150° at pressure of less than one micrometre, and then at 1,150°. The values ranged from 140 to 200 micrograms per gram. The most consistent values were between 140 to 160. The amount of carbon which could be converted into volatile carbon–hydrogen compounds was determined by pyrolling about 30 mg of the dust in an atmosphere of hydrogen and helium. The resulting involatile compounds were estimated by hydrogen flame ionization detector. The average value obtained was 40 micrograms per gram. Isotope measurements on the total sample gave us a C^{13} value of $+20$, relative to the PDB standard. These figures are considerably higher than those reported for intact meteorites.

Carbonaceous chondrites

For over a century, meteorites have been examined for the presence of organic compounds. The Alais meteorite was analyzed by Berzelius, the Kaba by Wholer, and Orgueil by Berthelot. It is now generally agreed that carbonaceous chondrites do contain polymeric organic matter. However, results obtained in the past have been ambiguous as to the origin of the detected extractable organic matter. Was it truly extraterrestrial in origin, or was it a result of terrestrial contamination? The analysis of the Murchison meteorite provided the first unambiguous and conclusive evidence for the presence of extraterrestrial organic compounds in meteorites. The Murchison Type-2 carbonaceous chondrite fell on 28 September 1969, near Murchison, Victoria, Australia. Several pieces were collected soon after the fall, and later during the months of February and March 1970. The stones selected for analysis were those with the fewest cracks, the least exterior contamination and of massive appearance. The samples used in the study contained 2% by weight of carbon and 0.16% weight of nitrogen. In these samples, aliphatic hydrocarbons were identified. Comparison with hydrocarbons produced from a spark discharge experiment showed a great similarity. Both appeared to be saturated alkanes containing the same dominant analogous series. These similarities suggest that the hydrocarbons in the Murchison meteorite may have originated abiogenically. Studies on the aromatic hydrocarbons reveal a similar result.

In the search for amino acids, interior pieces of the meteorite were pulverized, extracted with boiling water, and analysed by gas chromatography combined with mass spectrometry. A large number of amino acids were identified in the Murchison meteorite, including glycine, alanine, valine, proline, glutamic and aspartic acid. In addition there were a number of non-protein amino-acids. Of the amino acids with assymetric centers, both d and l enantiomers were detected. The presence of non-protein amino acids and amino acids with equal amounts of d and l enantiomers strongly suggest that these amino acids were prebiotic in origin. These provide us with the first conclusive evidence of extraterrestrial amino acids.

Studies similar to those undertaken on the Murchison meteorite have been applied to the Murray, which fell in Kentucky in 1951, and to the Mighei which fell in Odessa in 1986. These meteorites have been contaminated by exposure and by handling. The technique of gas chromatography and mass spectrometry enables us to separate the indigenous amino acids and hydrocarbons from terrestrial contamination. Thus far, three examples were found which showed amino acids of extraterrestrial origin.

More recently, the exploration of the Antarctic gave a new impetus to the study of meteorites. The Japanese geologist Kenzo Yanai on an expedition to the Yamato highlands of the Antarctic discovered eight pieces of meteorite outside his tent in 1971. When these meteorites were studied, he found they happened to be eight different meteorites. The following year, 300 were identified, in the subsequent year, 600, and now, over the years, 4000 new meteorites have been brought back from the Antarctic. Of these, about 40 fall into the category known as 'carbonaceous chondrites'. Two of these have been analyzed in the laboratory, the Alan Hills and the Yamato, and they provide supportive evidence of the observations made on the Murchison, the Murray, and the Mighei.

A striking feature of the Antarctic meteorites is that they are uncontaminated. The exterior and interior have the same amount of amino acids. Short of going to the asteriod belt, these are the best meteorites that are available to us for analysis. More recent work has established very clearly the presence of all five genetic bases in the Murchison meteorite.

Interstellar molecules

The presence of carbon together with hydrogen, nitrogen, and oxygen inevitably leads to the formation of carbon compounds. Astronomical observations confirm the presence in the galaxy of ammonia and water vapor. Since 1968, over 50 molecules have been identified in the interstellar medium. Several of these such as hydrogen cyanide, formaldehyde and cyanide acetylene are key molecules in chemical evolution and many have particular relevance to chemical evolution. There is even a suggestion that highly polymeric materials which may be present there are polyaromatic hydrocarbons. Others feel that these may be polysaccharides. Whether these individual molecules have contributed directly to the origin of life somewhere is probably difficult to visualize. However, it is intellectually very satisfying to see that the sequence from the atoms, hydrogen, carbon, nitrogen, oxygen to the small molecules, hydrogen cyanide, formaldehyde, appears to be a cosmic one. Further, to find the laboratory work in which molecules such as hydrogen cyanide and formaldehyde appear to be the intermediates in prebiological synthesis now being verified in the interstellar mediums adds a new dimension to our understanding of the process of cosmic evolution.

Conclusion

The studies which have started in the laboratory have taken us through our solar system to outer space. We are optimistic that the path of chemical evolution will be outlined in the laboratory. The biochemical knowledge which has been amassed within a few years has given us a deep insight into some of nature's most secret processes. With this understanding to help us, the time needed to solve our problems may not be long. We cannot deny the immensity of the prospect of man's philosophical position or shrink from its pursuit on account of the difficulty of the task, for after all what is required is to produce the simplest living entity.

12

The anthropic principle: self-selection as an adjunct to natural selection

BRANDON CARTER

1 Caveat Observator!

The purpose of this essay is not to review the rather copious literature (see e.g. Barrow and Tipler (1986)[1]) that has grown up on and around the subject of the anthropic principle and (and its sometimes rather far fetched extrapolations) since its original formulation as an explicit precept,[2] but rather to give a general introduction to the essential ideas involved before describing the particular application[3] that I consider most important.

I shall start at the very beginning by recalling that the *aim of science* (at least as I understand it), and hence the criterion by which the *scientific value* of any particular theory should be judged, is embodied in *Occam's principle* (or 'razor'), according to which the object of the game is to obtain a *logical* description of *as much as possible of what has been or may be observed* on the basis of *as few independent assumptions as possible*. In order to apply this principle – which can be summarised, using two key-words, as *comprehension* with *simplicity* – it is useful to respect two other subsidiary principles that are obvious enough so long as one retains an objective point of view, but which are often violated in contexts where subjective considerations prevail.

The first of these (as part of the requirement of simplicity) is what may be described as the *homology principle* to the effect that one should not introduce any unnecessary asymmetries. A particularly important special case that needs to be emphasised (precisely because it is specially prone to violation for obvious subjective reasons) is the generalisation of the *Copernican principle*, which was originally formulated as the precept that we should not necessarily suppose that *our own* earth is *central* to the surrounding planetary system. On a smaller scale this maxim extends to the *micro-Copernican principle* to the effect that one should not gratuitously presume that one's own social subunit (or oneself personally) is necessarily the focal heart of whatever smaller theatre of operations may be relevant. (In particular, Occam's principle rules against solipsism.) Similarly, on a larger scale, we have the *macro-Copernican principle* to the effect that we should not assume *a priori* that our own Milky Way is at the centre of the 'realm of the galaxies'.

The latter (unobjectionable and sober) extension of the Copernican principle is often replaced by a (questionable) teaching to the effect that the universe is actually *homogeneous* on a 'sufficiently large' (extragalactic) scale. Although this latter more extreme doctrine (which does in fact have a quite substantial basis of observational support) is commonly known as the Cosmological 'principle',[4] I would prefer to refer to it not as a 'principle' but merely as a 'law', to indicate its status as something that may well be empirically valid, but that might however need to be altered to take account of conceivable future observational information. In the present discussion the term 'principle' is to be understood as implying the higher constitutional status (albeit usually lower information content) of a precept whose validity is necessary *a priori*, as contrasted with the lower status of more detailed and specific laws whose forms might be postulated in various ways, between which a choice would have to be made *a posteriori* on an empirical basis. (The Copernican example could be demoted from the status of an unquestionably valid principle to that of an apparently valid (and indeed more informative) law simply by deleting the word 'necessarily' in its formulation above.) The unjustified elevation of the cosmological homogeneity property to the status of a 'principle' is an illustration of the way in which, since the time of Copernicus, the lesson embodied in what I have called the *homology principle* has been so well learnt by professional scientists (though perhaps not by people in other walks of life) that its application is liable to be taken unduly far – so as to tip the balance too much towards simplicity at the expense of comprehension in the application of the Occam principle.

In view of the preceding considerations, it became evident to me that it had become necessary to enunciate a second (complementary) subsidiary principle to restore the balance in favour of comprehension. This counterweight requirement is expressible in its most general form as what may be described as the *observational biassing principle* to the effect that one should not take the homology principle to the extreme of prejudging against the possibility of an intrinsic asymmetry between what is observed or observable and what is not. To put it succinctly (in the form of a traditional Latin motto), *caveat observator*! Although this is so obvious as to go without saying in many contexts (there is no need to remind a radio astronomer that his observations are biassed in favor of the detection of radio sources as compared e.g. with optical or X-ray sources) it is also prone to being overlooked in more emotive circumstances, which is what led me to insist on what I chose to designate by the term *anthropic principle* (by now too well established to be revisable, even if not the most ideally appropriate) covering the special kind of case for which the likelihood of bias arises not just from the characteristics of accessory measuring equipment (such as radio telescopes) but from the intrinsic nature and limitations of the scientist as a *human being*.

In so far as the very large (cosmological) scale observations I originally had in mind were concerned, the only relevant kind of bias would be that arising from selection effects, which as far as the anthropic aspect is concerned means more specifically the necessary restrictions on the conditions of time and place in which our scientific civilisation could have come into existence. This *macro-anthropic principle* is thus just the application to our own case of a more general *self-selection principle* which could (and should) be applied analogously by any other (extraterrestrial) civilisation engaged in scientific activity.

In the analysis of events on a smaller scale, purely passive selection effects might not be the only source of potential bias: in such a case the relevant *micro-anthropic principle* might have to include a reminder of the need to allow for causal effects resulting from one's own presence. Thus *whereas* the first astronomer (Hubble, actually) to observe a systematic tendency to recession of distant galaxies could plausibly apply (as he did) a Copernican type principle suggesting with appropriate humility that (as has been substantially confirmed by more recent observations) this represented a *homogeneous* expansion effect (rather than something specially related to our own position), *on the other hand* an ornithologist who noticed that birds flew more frequently away than towards him would clearly be excessively modest if he neglected the micro-anthropic principle by failing to recognise the likelihood that his own presence was (causally) responsible. In this hypothetical example the self-induced cause is too obvious to be at all likely to be overlooked in practice, but there are many actual examples of serious error resulting from failure to take account of the appropriate micro-anthropic principle in other more emotively charged circumstances: a typical example is that of the too-easily-flattered officer who underestimates the danger of mutiny by his subordinates through failing to take account of the extent to which he may himself have induced effects of both selective bias and amplification in receiving testimonies of loyalty.

Although it may perhaps have been noticed that in the last quoted example the observational biassing principle acted as an injunction to humility, there is not anything systematic about this: the oppositely directed homology principle was also effectively a precept of modesty in the preceding Copernican example. What is perhaps more systematic in either case is that there is in practice a greater danger of taking oneself too seriously than of not taking oneself seriously enough.

2   The anthropic principle in the case of Dirac's coincidence

The danger that can arise from a failure to be sufficiently objective is very well illustrated by the case that first drew my attention to this subject, namely an unjustified attempt by one of the greatest theoretical physicists of this century, Paul Dirac, to promote an unconventional theory according to which gravitational coupling strength is a variable function of cosmological epoch, using an argument based (only) on an observed order-of-magnitude coincidence of a certain 'cosmological large number' with a simple (negative) power of the gravitational coupling constant of the proton,

$$Gm_{\mathrm{P}}^2/hc \; \simeq \; 10^{-40},$$

where m_{P} is the proton mass, c the speed of light, and h and G respectively are what are traditionally known (in contexts where their variation is not envisaged) as the 'constants' of Planck and Newton. The reasoning in favor of time-variation was based on the fact that the cosmological large number in question was a dimensionless combination involving the Hubble age of the universe, t_{H}, as calculated by linear extrapolation back to the initial 'big bang' of the presently observed mean expansion rate of the distant galaxies. Although one would not expect it to be exactly the same as the actual time, t say, (as calculated after allowance for non-linear acceleration effects) since the big bang, one would nevertheless expect this simply calculated Hubble age to be of comparable order of magnitude (except in such exotic and no longer credible theories as that of the 'steady state' according to which the apparent 'big bang' never really occurred at all). Since this means that t_{H} should increase in approximate proportion to the true cosmological time t, Dirac argued that the gravitational coupling 'constant' would also have to vary in order for the observed numerical coincidence, which he considered too close to be fortuitous, to be maintained independently of cosmological epoch. The flaw in this line of reasoning is that it was dependent on the (in my opinion manifestly fallacious) assumption that there is nothing special about our own cosmological epoch. Being interested in the theory of stellar structure at the time I first learnt of Dirac's argument from the writing of Bondi,[4] I noticed that (as, I realised later, had recently been pointed out independently by Dicke) the relevant value of the Hubble age was of the same order as the theoretically calculable lifetime of a typical main sequence (hydrogen burning) star. It was evident furthermore that (as had also been pointed out by Dicke[5]) this was just what should have been expected within the framework of standard cosmological theory (with fixed gravitational coupling constant) since by that time (in the 1960s) it had become clear from the work of Fowler, Hoyle and others that all except the lightest (i.e.

hydrogen) of the elements of which our bodies are composed must have been made more recently than the big bang in (relatively rapidly burning) main sequence stars. Even at the earlier time in the 1930s) when Dirac introduced his hypothesis, it should have been apparent that it would be most likely for a life system maintained, as ours is, by a steady flux of energy from a nearby radiating star, to be formed before most such stars had burned out. The simultaneous application of these upper and lower anthropic age selection conditions leads to the expectation that if the conventional big bang theory is even roughly correct, the observable Hubble age at the time of our emergence should not differ too much in order of magnitude for the theoretically calculated lifetime τ_0 of a typical (hydrogen burning) star. Working this lifetime out in terms of the relevant microphysical parameters showed that (modulo relatively moderate adjustments involving much less extreme numbers such as the electromagnetic fine structure constant) the resulting expectation for the Hubble age is given by precisely the power of the gravitational coupling constant that corresponded to the observational relation whose significance had been guessed by Dirac. Although this showed that Dirac had not been mistaken in suspecting that this 'large number coincidence' was by no means fortuitous, its derivability in the framework of standard theory also showed that it was completely erroneous for him to have used this coincidence as a motivation for a radical departure from standard theory.

At the time when I originally noticed Dirac's error, I simply supposed that it had been due to an emotionally neutral oversight, easily explicable as being due to the very rudimentary state of general understanding of stellar evolution in the pioneering era of the 1930s, and that it was therefore likely to have been already recognised and corrected by its author. My motivation in bothering to formulate something that was (as I thought) so obvious as the anthropic principle in the form of an explicit precept, was partly provided by my later realisation that the source of such (patent) errors as that of Dirac was not limited to chance oversight or lack of information, but that it was also rooted in more deep seated emotional bias comparable with that responsible for early resistence to Darwinian ideas at the time of the 'apes or angels' debates in the last century. I became aware of this in Dirac's own case when I learned of his reaction when his attention was explicitly drawn to the 'anthropic' line of reasoning outlined above, on the occasion when it was first pointed out by Dicke in 1961. This reaction amounted to a straight refusal to accept the reasoning leading to Dicke's (in my opinion unassailable) conclusion that 'the statistical support for Dirac's cosmology is found to be missing'. The reason offered by Dirac is rather astonishing in the context of a modern scientific debate: after making an unsubstantiated (and superficially implausible) claim to the effect that in his own theory 'life need never end' his argument is summarised by the amazing statement that, in choosing between his own theory and the usual one (to quote his own words): 'I prefer the one that allows the possibility of endless life.' What I found astonishing here was of course the suggestion that such a preference could be relevant in such an argument, since there is of course nothing surprising about the preference as such. There are good Darwinian reasons for expecting an emotionally negative reaction to Keynes' law to the effect that 'in the long run,[6] we shall all be dead', and it is also understandable that the instinct for collective as well as individual survival should cause an unfavorable reaction to the recognition that there are also strong (though according to Dyson[7] not quite overwhelming) scientific reasons (involving the entropy principle commonly known as the 'second law' of thermodynamics) for believing that Keynes' law does not just apply microscopically to individuals, but that it generalises on a macroscopic scale to species, and even on a global cosmological scale to life as a whole. However, the mere fact that the relevant microscopic, macroscopic, or global version of Keynes' law may be displeasing is not a valid reason for discounting it in a scientific argument – nor for that matter (as any working actuary might warn) in planning a practical course of action.

The fact that such a distinguished veteran scientist as Dirac should have fallen into such an error of blatant wishful thinking is perhaps less astonishing when it is considered that his professional experience was mainly limited to purely theoretical aspects of the 'hard' physical sciences in which (in contrast with the situation in experimental work, and in the 'softer', e.g. biological and specially social sciences) it is possible, and very usual, to avoid thinking about problems of the *confinitive* type to be discussed in the next section. Pure theorists tend to prefer to concentrate on what I refer to as 'infinitive' problems of general or universal type, such as the (general) study of the properties of e.g. carbon in the solid state, rather than thinking about the kind of everyday 'confinitive' problems that an experimentalist might have to face in making practical decisions in particular cases, (such as whether an atypical result from an expensive (and therefore in practice unrepeatable) destructive experiment on a particular diamond crystal should be discounted because of conceivable impurity effects). If one's entire working experience has been that of a theorist concerned only with the study of 'infinitive' problems, it is perhaps not surprising that one should find oneself mentally unprepared when finally forced to face up to confinitive problems in the context of cosmology. Dirac's error provides a salutary warning that contributes to the motivation for careful formulation of the anthropic and other related principles.

One such related issue is the notion that theories can only be refuted but never decisively confirmed, which may sometimes make some sense for theories of infinitive type, aimed at universality, but which may be very misleading with respect to theories of confinitive type aimed more modestly at partial understanding of particular situations, for which there is no such fundamental asymmetry between refutation and confirmation. (In choosing between a theory based on the hypothesis that an unseen coin fell head up and an alternative theory postulating that it fell head down, refutation of one may be equivalent to confirmation of the other.) Even for theories of infinitive type the idea that refutation can be decisive is potentially misleading in view of the fact that the interpretation of a purported experimental refutation can never be entirely independent of confinitive questions. (One might think naively that any theory based on the postulate that space-time has more than four dimensions could be refuted at once on the basis of everyday experience, but modern theoreticians are uninhibited in taking such theories very seriously indeed, knowing that the chain of inference between observation and theory is so long and complicated that *prima facie* refutation may be as illusory as that more notorious will-o-the-wisp *ultimate fundamentality*.)

3   Treatment of scientifically confinitive systems

The most favorable situations for scientific study are those in which one has the advantage of dealing with what I refer to as a *scientifically infinitive* system, meaning a system of a generic kind in which there is no (effective) limit to the number of possible independent tests of theoretical predictions either by artificial experiments (in fields such as laboratory chemistry) or observations of naturally occurring phenomena (in fields such as the astronomical study of stellar atmospheres). If (as is typically the case in many fields of purely theoretical science) one's training and experience has been limited to such favorable situations, one may be ill-prepared (and perhaps psychologically at a disadvantage even compared with an untrained person) in the alternative situation where one is faced with the problem of dealing with a system that is *scientifically confinitive* in the sense that the information available in practice or in principle is limited so that beyond a certain point no further direct experimental or observational input can be expected. In such a case however, one's theoretical training should at least be helpful in suggesting the possibility of trying to escape at least partially from such limitations, so as to obtain additional indirect information, by imbedding the confinitive system under consideration in a larger infinitive system, and accepting that this will typically entail treatment at a less detailed level of resolution.

The kind of scientifically confinitive situation in which a theorist might be interested typically arises in the study of a unique and unrepeatable event. A classic example of a scientifically confinitive system is the French Revolution, whose causes have been the subject of much discussion and analysis, though not of any clear consensus. The typical first (unconstructive) reaction of a scientist faced with the manifestly speculative (albeit often dogmatically expressed) suggestions put forward by writers as diverse as A. de Tocqueville and K. Marx, is to lament the impossibility of subjecting the system to experimentally controlled changes in the 18th century background conditions. The next (more constructive) reaction is likely to be to follow the example of some of these writers in attempting to extend the range of study to the less confinitive field consisting of *all* political revolutions in human society – which in practice means restricting oneself to rather vague generalities and (if one is honest) giving up the illusory hope of a really full and detailed understanding on any objectively firm basis.

The ultimate example of a scientifically confinitive field of study is of course cosmology – the study of the universe as a whole – in which the situation is in some ways even more frustrating. Although (again at the expense of vagueness) one may succumb (as I have previously done myself when introducing the 'strong anthropic principle'[2]) to the temptation of seeking a more comprehensive insight at a purely *theoretical* level by imbedding the confinitive system consisting of our own universe in an infinitive system consisting of a conceptual ensemble of universes, one cannot thereby gain any additional information of an *experimental* or *observational* nature.

While it may indeed be a serious drawback, the impossibility of subjecting theory to the test of experiment or application to independent examples does *not* mean that the study of cosmology or of other scientifically confinitive systems (such as the French Revolution) is entirely beyond the scope of scientific method. What it does mean is that one should be more than usually careful in order to extract the maximum understanding from the limited information available, while at the same time avoiding the pitfalls involving overhasty conclusions from flimsy evidence, particularly on the basis of various kinds of subconscious bias (that might have been innocuous if subject to experimental filtration).

The meaning of some of the foregoing remarks may be illustrated by considering a rather simple hypothetical example of a scientifically confinitive problem of a familiar kind on a parochial scale. Consider the problem posed at the trial by a local jury of a travelling salesman who, unlike everyone else in the small town through which he happened to be passing, unfortunately had no (apparent) alibi at the time a murder was committed there, no other relevant direct evidence being available. Before making a practical decision (which might be *deliberately* biassed, e.g. by a legitimate human rights presumption of innocence, or by a natural but less legitimate desire to satisfy the local mayor's need to obtain a convenient scapegoat) the jury would (presumably?) wish to form some idea of what was most plausible from a *scientifically objective* point of view. At this stage in the process, one might invoke indirect evidence of innocence by situating the (scientifically confinitive) problem of the particular murder under consideration in the broader (scientifically infinitive) context of murders in general, arguing for innocence on the statistical ground that it is observationally known to be comparatively rare for the culprit to be a stranger to the victim, but comparatively common for alibis to be faked by collusion between interested parties. For our present purpose it is more instructive to consider the alternative possibility that such statistical information might not actually be available on any solid factual basis – so that as far as *observational* evidence is concerned the attempt to escape from a confinitive to an infinitive system would have failed (as it must always do in the case of cosmology) – but that the corresponding information might nevertheless be obtained approximately from purely theoretical plausibility arguments. Although better than nothing, such theoretical information should of course be subject to caution in the absence of observational confirmation: the type of danger that might arise is illustrated by the possibility that, faced with a defence based on such theoretical plausibility considerations, the prosecution might correspondingly be tempted to exploit the natural tendency to xenophobia that is endemic in virtually all social units, by arguing that it was also plausible to suppose that outsiders were more prone to crime than local townspeople. An insufficiently alert jury might fail to appreciate that a purely theoretical argument on such lines can be countered (without resort to observation, which might tip the balance either way) merely by the *homology principle* consideration (in effect a micro-Copernican principle according to which one should not assume gratuitously that one's own home town is at the centre of the universe) that one man's foreigner is another man's native and vice versa.

Before leaving this example it is to be remarked that there is nothing unique about the way in which the confinitive system under consideration may be imbedded in a more extended system: instead of situating our murder case within the class of all murder cases anywhere, we might situate it within the class of all events in the history of the small town: this might make it possible to strengthen the prosecution case by attaching statistical significance to the *temporal coincidence* of the salesman's visit with the occurrence of the crime *if* (but only if) it could be demonstrated that visits of salesmen were too rare for such a coincidence to be likely as a fortuitous accident. This latter eventuality provides another example of the risk in drawing too hasty conclusions from what may indeed be a significant coincidence: in the alternative theory, of a local conspiracy with faked alibis, the coincidence might also be explicable as being due to timing deliberately chosen for the purpose of framing an outsider. The purpose of going through this gedanken[9] exercise (whose outcome is evidently a stalemate between rather weak direct evidence of guilt and comparably weak indirect evidence of innocence, which – except in Scotland where a scientifically sound 'not-proven' verdict would be legally admissible – would oblige the jury to make their decision on moral, pragmatic, or other non-scientific grounds) is not to arrive at particular scientific (still less judicial) conclusion, but to provide a simple illustration – addressed particularly to readers less familiar with detective problems than with the more tractable scientific problems of infinitive type – of the possible methods (and associated dangers) of dealing scientifically with systems of confinitive type.

Before going on to discuss a problem that (unlike the essentially pedagogical examples mentioned so far) I consider to be of major scientific interest in its own right, I would like to pause to emphasise the importance of avoiding a dangerously blinkered notion of what constitutes the scientific method. In particular I would like to contribute to clearing up the widespread misunderstanding that has lead to undiscriminating insistence on the requirement that a theory should satisfy the requirement of 'refutability' (originally motivated as a healthy reaction against empty and arbitrary assertions in dogmatic ideologies) which may sometimes be quite inappropriate, specially in dealing with systems that are confinitive in the sense introduced above. What is traditionally known as *the* scientific method of justifying a theoretical explanation of a known system of facts consists of using the theory for the derivation of further independent *predictions* that can be experimentally or observationally verified. However although the possibility of making such applications may indeed provide the practical justification for putting effort into the construction of such a theory, it is nevertheless a mistake (albeit a common one) to suppose that the possibility of making independently verifiable predictions (which does not arise at all in the example given above) constitutes the only scientifically valid method of justifying a theory. In so far as scientific status (as opposed to practical utility) is concerned, the logical absurdity of systematic discrimination against theories that are 'non-refutable' in the sense that they do not provide independently verifiable predictions is evident from the consideration that yesterday's independent predictions will (if successfully verified) be part of today's system of previously known facts. When a logical consequence of a theory ceases to be a (refutable) prediction by becoming an (empirically confirmed) 'postdiction' the theory clearly becomes not less but more satisfactory from the point of view of pure scientific understanding. The appropriate criterion for the scientific value of a theory is Occam's requirement of maximum (relevant, empirically valid) *deductive output* from minimum (independent) *inductive input* (what may be debatable being how to weigh and judge relevance, independence, and empirical validity of the input and output information). In so far as 'refutability' means 'verifiable predictive output' it is certainly better than unverifiable output and much better than non-deductive assertions (not to mention deductions that are empirically false) but, far from being indispensible, such 'refutability' is definitely less satisfactory (scientifically) than an equal amount of (irrefutable) 'verified postdictive output' provided the latter is deduced logically from the same amount of independently hypothesised input information.

4   Global view of our biosphere as a confinitive system

What is by many criteria the most important example of a scientifically confinitive system (intermediate between the scale of the French Revolution and that of the universe as a whole) is provided by the *terrestrial life system* or 'biosphere' whose evolution has 'culminated' (to use a rather subjective expression) in the emergence of the scientific civilisation to which readers of this article may be presumed to belong. The effort put into the study of details of this system by various categories of specialists such as botanists, biochemists, palaeontologists etc. (not to mention historians of the French Revolution) is at least comparable with that put into all other scientific activities taken together, and the wealth of resulting knowledge is proportionate. (Many particular aspects, such as biochemistry, are in effect scientifically infinitive, and thus amenable to the method of repeated experiment). Despite all of this, however, our global understanding of the system as a whole is limited to a few rudimentary fragments (among which, of course, the simple Darwin–Wallace notion of natural selection has pride of place) and it is likely to remain so unless and until we gain access to other comparable life-systems in the universe. In the meantime (which, I have strong reason to believe, means throughout the foreseeable future) we have to cope as best we can with the difficulty that as far as its global evolution is concerned our 'biosphere' is effectively a confinitive system in the sense of the previous section. Although (for the time being) we cannot empirically observe (still less tamper experimentally with) other analogous life-systems, this confinitive situation does not prevent us from trying to improve our understanding by thinking about such possibilities on a purely theoretical basis (in the manner illustrated in the previous section for small scale examples of confinitive systems).

There is in fact more than one potentially interesting way in which the (confinitive) terrestrial life-system might instructively be considered as a subsystem of a more general (infinitive) system: it might be considered as a particular case within the class of all planetary systems (with or without any form of life) or as a particular case within the class of all life systems (whether or not of planetary type). In the present investigation we shall be concerned essentially with the first of these possibilities.

Although present day understanding of the mechanisms involved in the formation of a planetary system like ours could at best be described as sketchy, there seems to be a general consensus to the effect that such systems are likely to be of common occurrence as a bi-product of the normal pro-cesses of stellar formation. Moreover, although the particular kind of geophysical and chemical conditions that originally prevailed on our own planet may not have been of the most common occurring type, on the other hand there are no very strong reason for believing such conditions to have been exceptionally rare either. This widely agreed consensus has been the point of departure for a lively (and sometimes even passionate) debate between supporters of the conclusion that life-systems such as ours should be of comparatively common occurrence in stellar systems, and those (including the present author) who think that the balance of evidence is weighted against this conclusion. To arrive at such a conclusion it would of course be sufficient to justify the hypothesis to the effect that the life-systems under discussion must originate automatically whenever a favorable geophysical and chemical environment occurs, but the validity of this last hypothesis is the point at which all concensus breaks down in the absence of any sufficiently detailed theory of the mechanisms of the origin of life, there being (at present) no sufficiently detailed description of the relevant mechanisms to provide even the crudest estimates of relevant basic quantities such as characteristic time-scales. Conversely, if it could be established that life-systems were common by other (e.g. observational) means then one would have an important piece of information restricting plausible theoretical models for such mechanisms. It is in this connection that supporters of the conjecture that life is common are liable to fall into the trap of overlooking the *anthropic* principle by arguing that since life is observed to occur on at least one out of a collection of a dozen or so planetary bodies in a system associated with an apparently typical star, therefore life might be expected to be proportionally common in other planetary systems. The flaw in this reasoning lies in the uncritical (usually subconscious) application of the homology principle to justify the argument that, in the absence of any observable differences between our sun and the many other stars of the same spectral type, its planetary system should not be assumed to be untypical. The point is that the absence of positively observed asymmetry is not incompatible with actual asymmetry between what is observable and what is not. Moreover (as the anthropic principle reminds us) there is an obvious asymmetrical self-selection effect ensuring that our own planetary system (which, subject to present day technological limitations, is the only one for which the relevant observations of life can be made at all) necessarily belongs itself to the subclass containing life – no matter how small a fraction of the total this subclass may be. The upshot is that taken by itself the mere existence of life on our own planet can provide no evidence at all in favour of, or against, the conjecture that life is common on similar planets elsewhere.

Apart from protecting us from an elementary but common error, this particular application of the anthropic principle has not been much use in providing any information effectively restricting plausible theoretical models of the origin or evolution of life. We can however obtain more interesting and restrictive conclusions – including the *prediction of falsity of the conjecture that life is common* in other planetary systems – from the rather more elaborate application of the anthropic principle that will be described in the following section. (This prediction has been put forward as a counterexample to the suggestion that the anthropic principle is unable to provide observationally verifiable, and thus potentially 'refutable' results, but the importance of this should not be exaggerated since, as was insisted in Section 3, 'refutability' in this sense is by *no* means indispensible for scientific respectability.)

5   Self-selection as an adjunct to natural selection

Although the 'apes or angels' debate has long ago been settled in favour of the former (at least in so far as the scientifically educated part of the community is concerned) in the sense that there is general concensus in accepting the existence of evolution as sketched out roughly by the fossil records, there is however still much discussion as to whether the Darwinian mechanism of natural selection provides a sufficient explanation. There have in fact been so many attempts to cast doubt on the established body of Darwinian theory that its advocates (with whom I would class myself) have often felt obliged to describe themselves as neo-Darwinians in order to emphasise that the body of theory to which they subscribe is now much more complete than that originally put forward by Darwin and Wallace, in so far as it incorporates modern understanding of Mendelian genetics in general, and of the genetic code and of mechanisms of mutation in particular. Many of the more technical reasons for raising doubts (such as, for example, the frequent appearance of jerkiness in rates of evolution) can and have been explained without difficulty, at least in principle, within this reinforced framework of standard Darwinian theory. Nevertheless there remain doubts and objections of a more fundamental kind, which have been less adequately treated in the professional literature on the subject, and which I wish to deal with directly in the present section.

The doubts and objections in question are based on the widespread impression that the long term tendencies of evolution as deduced from the fossil records are far too systematic to be fully explicable on the basis of such an essentially aleatory mechanism as pure Darwinian natural selection: they seem to be *purposefully directed* towards the emergence of humanity. Much of this appearance is of course mere illusion resulting from naive preconceptions, as can be seen from the once fashionable suggestion that the horse had been created on purpose to provide means of transport for man. Whereas such a proposition as that would merely raise smiles if put forward in educated circles today, on the other hand, no-one could seriously dispute that it is reasonable to account for the horse's *own* legs as having been designed for the purpose of transporting *itself* — such an account being of course fully consistent with a causal explanation in Darwinian terms. (Examples such as the latter illustrate the way in which a mechanism that is of the ordinary causal type when considered over the long term can give rise to shorter term effects that are most conveniently described using teleological terminology.) The most serious arguments in favour of the existence of a mysterious supra-Darwinian guiding influence to explain the observed course of evolution are based on examples that are neither so obviously Darwinian as the development of the leg as an instrument for displacement nor so obviously absurd as the idea of the predestination of the horse for human transport, but on a general impression (too vague to be either substantiated or invalidated on a rigorous basis) that the observed tendency towards the successive development of life forms that are progressively more 'advanced' ('advancement' being understood as defined with respect to the standards set by our own species) is too systematic to be explicable by the blind action of ordinary Darwinian selection on the basis of circumstantial requirements for short run survival.

I do not wish to offer any judgement on whether this widely felt impression of purposeful direction in evolution can actually be considered to be an intrinsically sound assessment (in an area where there is a strong danger of illusion arising from preconceptions). What I do wish to point out, however, is that (whether or not it is in practice observable) *some such appearance of purpose is in principle entirely predictable* (without invoking any mysterious design mechanism) provided one bears in mind that in addition to the effects of (Darwinian) natural selection one should also take account of the effect of (anthropic) *self-selection.*

Just as the teleological development of an organ during the growth of an individual is causally explicable in terms of the action of natural selection on the species, so also, on a much larger scale, one would expect on the basis of the self-selection principle that the evolution of our life system should manifest an apparently teleological tendency to systematic advancement towards our own state of development, *whether or not* such a tendency were typical of the result of the stochastic interplay of Darwinian evolution mechanisms in general. It is perfectly conceivable (and in my opinion, for reasons to be explained in the next section, most likely) that the *typical* result of the evolution of life-systems on other planets elsewhere in the universe should be very different from what is observed here, with no systematic tendency to 'advancement' and no 'intelligent observers' or 'scientific civilisation' as the outcome. Nevertheless, in so much as there has been (more or less) progressive advancement up to our own stage of development here (and in so much as our present stage is not yet so advanced as to enable us to detect other comparably advanced civilisations at the distance of even the nearest stars), this 'progressive advancement' certainly need not be supposed to have been inexorable, and therefore cannot be used as an argument in favour of the importance of non-Darwinian evolution forces. The fact that we do not observe the type of alternative outcome for which, in the absence of 'progressive advancement', no 'intelligent observers' ultimately emerge, is obviously of *no significance* as evidence against the probability of such an outcome, for the simple reason that such an alternative outcome is outside our (present) scope for observation. No matter how intrinsically improbable the combination of chance developments that may be necessary for the emergence of intelligent observers, the self-selection principle almost tautologically excludes the possibility of our finding ourselves to have emerged anywhere that such a combination failed to occur. Once it is appreciated that our own case may well, for this 'anthropic' reason, be highly untypical of what would be most commonly be produced by the Darwinian evolution mechanism, it can be seen that there is *no longer any basis* for the popular arguments to the effect that the Darwinian mechanism needs to be supplemented by some mysterious (external?) evolutionary guiding force in order to account for what appears to have occurred on earth.

The outcome of the preceding reasoning is somewhat analogous to that of the example involving the gravitational coupling constant that was presented in Section 2. In both cases taking proper account of the anthropic principle is shown to destroy the purported empirical basis for radical (I am tempted to say heretical) departures from the main body of established theory (as associated with the respective names of Darwin and Newton) for underlying motives which (as I showed explicitly in the previous example, and as is obvious in the present case) may be far from being purely scientific. However it should not be supposed that the effect of the anthropic principles can only be to reinforce (scientific) conservatism: I shall describe in the next section the way in which, when applied in a somewhat more detailed investigation of the terrestrial life-system, the anthropic principle led me to conclusions that were startlingly different from my prior preconceptions.

6 The significance of the solar-age coincidence

Having shown (by anthropic considerations) why certain superficially striking appearances should be considered as being less significant than is commonly supposed, I should now like to draw attention to an observational coincidence that I believe (again for anthropic reasons) to be much more significant than has yet been generally appreciated. The coincidence in question is the observed agreement in order of magnitude (within a factor of about two, i.e. with much greater precision than the example discussed in Section 2) between the present age of our terrestrial life-system (nearly 5×10^9 years) and the theoretically estimated lifetime (of the order of 10^{10} years) during which the sun can continue to support itself by steady hydrogen burning as an ordinary main sequence star.

What this means is that our arrival at the present stage of development was achieved by a rather narrow squeak – a 'close run' – at least on a logarithmic timescale calibrated with respect to the enormous scale of our uncertainty concerning the mechanisms controlling natural rates of evolution: if our biological evolution had proceeded more slowly by a mere factor of two or so, as compared with the rate of thermonuclear evolution of the sun, we would never have come into existence at all before it was too late. Such a conjunction has precisely the hallmark of an anthropic selection effect – while being patently inexplicable (except, implausibly, as a mere fluke) by any conceivable alternative mechanism of a more direct nature. Assuming that in a typical case the rate of thermonuclear evolution of a star is quite distinct from the statistically expected rates of evolution of 'life' on any associated planets, the obvious induction is that it must be the latter whose order of magnitude is longer, and that the self-selection effect by which our own history may be presumed to have been atypical (as was discussed in the previous section) includes the requirement of exceptionally rapid evolution, so as to reach the stage of emergence of 'observers' within the limited available time. Just as the ordinary Darwinian selection mechanism is expressible by the catchphrase 'survival of the fittest' so the supplementary self-selection mechanism involved here might be expressed as 'arrival of the quickest'!

The lack of earlier recognition of the significance of the solar-age coincidence would seem to be largely attributable to failure to recall the usual methods of dealing with confinitive problems on a parochial scale (as described in Section 3) when confronted with a confinitive system on the more awe-inspiring scale of the (entire history of) the terrestrial life system. Our earlier discussion (in Section 3) was intended as a reminder of the more comprehensive understanding that can be gained by situating an individual (confinitive) system in the wider context of an (infinitive) extended system, as an element of an appropriately defined class – even if only theoretical information is available about the other members (and of course all the more so if observational information is also available). The elucidation of the solar-age coincidence as described above illustrates how it is instructive to consider our terrestrial life system as a particular example (at present the only observationally accessible one) in the general class of planetary based life-systems, this class being itself a subset of the class of *all* planetary-environmental systems (with or without life). I shall now describe how this method of analysis can be carried considerably further.

The discussion in the previous section raised the question of the extent to which our particular history can be regarded as typical within the class of histories of planetary environments in general: what, to use the terminology of Monod[8], was the relative importance of chance as opposed to necessity? Now, provided the genetic mutation rate and the number of individuals in a species (or in a smaller sexually interacting population) are not too small (in which case aleatory non-Darwinian evolutionary drift may be important), the standard theory[9] of Darwinian selection predicts continuous adaptation to the environment on timescales corresponding to only a moderately large number of generations. Although such short term Darwinian evolution of a single species or smaller population subunit is essentially deterministic (or 'necessary' as Monod would put it) any practicable (not too unreasonably detailed) scheme of description of the long term evolution of the enormous number of ecologically interacting species making up the life-system as a whole will nevertheless be essentially *stochastic*, i.e. dominated by 'chance'. Were such a description available, it would, as an end result, provide an intrinsic probability distribution (dependent on overall geophysical background conditions) for the time, t_0 say, of emergence of 'intelligent observers' such as ourselves. At a more detailed level it would also provide probability distributions for the times (t_1, t_2, ... say) of attainment of other less 'advanced' stages of development, such as the emergence at the most primitive level of 'life' itself (whatever that may be defined to mean exactly), that are presumably prerequisite as intermediate stages on the way.

Now although it is quite beyond the scope of present-day science to provide such a description in any complete and quantitative form at even the crudest level, I wish to demonstrate nevertheless that it is possible (using the anthropic principle and the coincidence stated above) to obtain some interesting restrictions on the most likely form of the *intrinsic* probability distribution that such a description would provide, as measured against the *externally* imposed timescale, τ_0 say, during which the requisite (geophysical background) conditions are maintained – the main sequence lifetime of the sun being a plausible first guess, and certainly an upperlimit on the relevant value of τ_0 in our own case. To start with, we can roughly classify the steps between intermediate stages of 'advancement' into two categories, namely 'easy' steps, for which the mean time required for passage to the next necessary stage is small compared with the time τ_0 available, and 'difficult' steps for which the intrinsically expected time is large compared with the extrinsically available time. For the purposes of the present discussion, in which we shall not make any attempt at time resolution on scales small compared with τ_0, we can consider the 'easy' steps as being effectively deterministic, in so much as they will be traversed with probability close to unity. As far as a crude probabilistic model is concerned we can therefore leave the (evidently very numerous) 'easy' steps out of account, and restrict our attention to the comparatively small number, n say, of essential steps (which might for example include the original development of the genetic code) that are 'difficult', i.e. that have a low probability of achievement within the available time.

Now so long as it remains low compared with unity, the total probability P of realisation of a single event dependent on the random occurrence of an appropriate combination of (in our context, ecological) circumstances will grow approximately linearly as a function of time t. It follows that the probability of realisation of two such events in succession will therefore grow quadratically with time, and hence by induction it can be seen that the probability that a total of n such events will have been realised as successive steps within a given total time t will be given approximately by a power law expression of the form

$$P \simeq \alpha t^n$$

for some fixed coefficient α which, in the present application must be sufficiently small to ensure that P remains substantially less than unity throughout the available time range,

$$0 < t < \tau_0$$

The foregoing formula, valid for $n \geq 1$, is to be contrasted with the kind of probability distribution that would apply if there were no 'critical' (i.e. essential and 'difficult') steps at all: if all the (essential) steps were sufficiently easy P would be approximately a Heavyside function, in that it would rise from zero to close to unity after a time t comparatively small compared with τ_0. The corresponding differential probability distribution for the time of achievement of the nth 'critical' step, will be given by

$$dP/dt \simeq n\alpha t^{n-1},$$

This last expression will be approximately valid not only for $n \geq 1$, but even over the greater part of the time range. In the 'easy' case $n = 0$, for which the non-zero contribution will be concentrated as an approximate Dirac delta function near the origin, $t \simeq 0$. At the other extreme, if n were very large, the distribution would be concentrated very near the upper limit of the time range, at $t \simeq \tau_0$. Only in the intermediate case, $n \simeq 1$, will the probability be distributed *uniformly* over the entire available time range. These tendencies are faithfully reflected in the expression for the statistically expected value, t_e say – as defined relative to the (*anthropicly*) restricted distribution specified by the extrinsic cut off, $t \leqslant \tau_0$, of the later part of the time range – for the emergence of the postulated 'intelligent observers'. At the present level of approximation this expectation value will be given for any supposed value n (including $n = 0$) of the prerequisite number of 'critical' steps by

$$t_e/\tau_0 \simeq n/(n + 1).$$

This basic (anthropic) relation between the three quantities n, τ_0, t_e can now be used to *predict* the value of any one of them provided we already know the values of the other two. If (in the absence of any suggestion to the contrary) we suppose that the actual measured age of the earth provides a fair estimate of the expectation value t_e, and if we assume that the calculated main sequence lifetime of the sun gives the appropriate value of the time τ_0 that was available for biological evolution, then we can interpret the coincidence referred to at the beginning of this section as signifying the likelihood that the number of 'critical' steps is not less than nor very much more than unity, i.e. that we shall have

$$n \simeq 1.$$

To be more explicit, we can very convincingly rule out the possibility that n should be zero, because it is extremely hard to think of any plausible reason why our arrival should have been exceptionally delayed with respect to what was probable *a priori*. This has the corresponding implication that (contrary to what many science fiction writers would have us believe) 'intelligent' life must be *rare* even on planets with favorable environments, occurring only where chance has brought about exceptionally rapid evolution. At the other extreme the fact that (before swelling so as to engulf the earth in fire) the sun, although no longer young, has yet a substantial fraction of its main sequence life still to run is incompatible with what one would expect if n were large (more than two or perhaps three) compared with unity – *unless* there is some other (less obvious) reason why the development of terrestrial life should be cut off at an earlier stage, long before the end of the main sequence phase. We shall briefly consider the plausibility of this latter alternative interpretation in the next section.

7   How close a run?

When I first realised the implications that I have just described I was rather surprised, because my exposure to the arguments put forward by the defenders of the idea of purpose in evolution had left me with the impression that the number n of 'critical' steps whose passage appeared to have been intrinsically improbable *a priori* (albeit anthropicly inevitable *a posteriori*) was quite *large* compared with unity, may be in the range between ten and a hundred. Furthermore a more extreme proposition to the effect that the correct number of effectively 'critical' steps is enormously larger even than this has recently been advocated by Barrow and Tipler,[1] who have insisted on the importance of the loophole mentioned at the end of the derivation of the *prima facie* upper limit obtained above. After providing an explicit catalogue (going considerably beyond my original tentative short list of candidates) of the kind of developments that might appear to qualify for the status of 'critical' steps, Barrow and Tipler go on to discuss my willingness to concede that most such steps must have been much less 'difficult' (or else less 'critical' in the sense of being anthropicly necessary) than I had imagined: they suggest that I might have been being unduly conservative in overhastily dismissing the possibility that the sensitivity of the terrestrial climate to small perturbations might have predestined our biosphere to destruction in the comparatively near future as a result of a quite small increase in the thermal output of the sun.

I should like to take this opportunity of replying firstly that I agree with their assessment that our understanding of the stability of the climate is insufficient to allow the exclusion of such a possibility – the prediction of such a cut-off being a viable possibility as an alternative to the conclusion that n is very small. However I would also like to express my reservations with respect to their advocation of the extreme proposition that they refer to as a 'more radical approach' to the estimation of n, in which they go so far as to suggest that n might be comparable with, or even that it might exceed, the number of human genes, estimated by Dobzansky as of the order of 10^5 – which would be compatible with the formula in Section 6 only if the earth's climate had only a few tens of thousands of years stable existence still ahead of it at the time of our arrival.

My reason for resolutely disagreeing with such a 'radically' high *a priori* estimate of the likely value of n, which is tantamount to the supposition that virtually every minor step in our evolution was 'critical' (meaning that it was both essentially *necessary* and 'difficult' in the sense defined above) is that it would seem to be based not only on a gross underestimate of the number of alternative ways of achieving similar results (as evidenced by the rich diversity of known life-forms) but also on a lack of recognition (amounting almost to a complete denial) of the importance of the Darwinian selection mechanism whose efficiency in achieving continuous progress has been amply demonstrated by both theoretical and observational means in numerous examples. I find it difficult to doubt that this mechanism ensured that the passage through nearly all of those minor steps in our evolution that were necessary at all (i.e. not merely redundant as being of neutral survival value) was 'easy' in the sense defined above. The value of n is bounded above by the number of evolutionary path segments that are favored by natural selection on a short term basis, a number that is certainly very small compared with the total number of microscopic genetic changes involved. The possibility of 'difficulty' at certain crucial steps arises from the fact that such 'easy' highways of continuous Darwinian evolution may not necessarily lead consistently in the direction of long term survival advantage (still less of 'progress' in our subjective sense of the word), and that they may even be blocked by dead ends beyond which long term gains would be achievable only at the expense of short term sacrifices. In order to link the easily followed path segments together to form the required globally complete evolutionary path or tree, it may be necessary on occasion to wait until an exceptional combination of ecological circumstances (analogous to the displacement of a watershed by an earthquake) temporarily changes the favored direction in the necessary sense, or until other processes (such as random genetic drift in an isolated small population) have made it possible, with sufficient luck, to jump across a potentially 'difficult' step by non-Darwinian means. However, although I was inclined (before taking account of the solar-age coincidence) to believe that the number of such critical steps might be very large compared with unity, on the other hand it has always seemed clear to me that it must be extremely small compared with the total number of bits of information involved, or even compared with the more moderate intermediate number of gene units, since in so far as most minor steps (on the scale of a single gene or less) are concerned I would expect short term and long term advantage to be compatible. In other words, on the basis of the fairly obvious preponderance of continuous and effectively 'easy' development (by locally deterministic Darwinian natural selection, or by random genetic drift) at nearly all stages of our evolutionary history (as evidenced in numerous flourishing side branches that are clearly irrelevant to our own existence) I would confidently guess that the number n of 'difficult' gaps (to be crossed more by 'chance' than 'necessity') while perhaps large compared with unity (as I originally supposed) will nevertheless be extremely small compared with the number of microscopic gene changes involved.

Although I must admit to being unable to translate the foregoing 'neo-Darwinian' considerations into a quantitative demonstration that the 'critical' step number n cannot plausibly be anywhere near as large as the gene number, nevertheless, far from ruling out large values of n on the basis of any prejudice in favour of long term stability of the climate, I would on the contrary prefer to reason in the converse sense, by using the above 'neo-Darwinian' line of argument on conjunction with the anthropic formula of the previous section as evidence of reasonably long term stability of the climate (against the effect of a slightly hotter sun). Unfortunately, this (not quite rigorous) reassurance that (contrary to the rather alarming suggestion of the 'radical approach') the earth's climatic environment could after all survive for substantially more than a few times 10^4 years (if not necessarily for the whole 10^{10} year main sequence lifetime of the sun) is a guarantee only against the possibility of external destabilization (by the at present slow but steady heating of the sun): I would emphatically warn that it does not offer any assurance whatsoever against *intrinsic* destabilization (e.g. by industrial atmospheric pollution) resulting from the action of our own civilization.

Acknowledgements

I should like to thank John Wheeler for his original encouragement to write on this subject, and to thank John Barrow and Frank Tipler for recent discussions and correspondence.

References

[1] Barrow, J.D. and Tipler, F.J. (1986). *The Anthropic Cosmological Principle*, Clarendon Press, Oxford.
[2] Carter, B. (1974). Large number coincidences and the anthropic principle in cosmology, in *Proc. I.A.U. Symposium*, **63**: *Confrontation of Cosmological Theories with Observational Data* (ed. M.S. Longair), Reidel, Dordrecht.
[3] Carter, B. (1983). The anthropic principle and its implications for biological evolution, *Phil. Trans. Roy. Soc. Lond.*, **A310**, 347.
[4] Bondi, H. (1960). *Cosmology*, Cambridge University Press.
[5] Dicke, R.H. (1961). Dirac's cosmology and Mach's principle, *Nature*, **192**, 440.
[6] It may be that I have omitted the (quite unnecessarily restrictive) interjection '... gentlemen ...' at this point in the folklore version of the Keynes quotation.
[7] Dyson, F.J. (1979). Time without end: physics and biology in an open universe, *Rev. Mod. Phys.*, **51**, 447.
[8] Monod, J. (1970). *Le Hazard et la necissité*, Editions du Seuil, Paris.
[9] A convenient mathematically oriented introduction to evolution theory is given e.g. by J. Maynard Smith, *The Evolution of Sex*, Cambridge University Press (1976).

13

Astrology and science: an examination of the evidence

IVAN W. KELLY, ROGER CULVER, PETER J. LOPTSON

Astrology is the theory that particular configurations of the heavenly bodies at one's birth are significantly related to one's personality or character and life events at adulthood. Astrology claims to be able to describe and predict a person's behavior on the basis of a horoscope. An individual's horoscope is a geocentric map of the sky in which are located astrologically significant elements – such as zodiac signs, planets, the ascendant, and the houses – at the time of that individual's birth.

Astrologers visualize the universe as a giant celestial sphere with the earth as its center. From this perspective the sun and planets appear to move around the earth on a yearly basis. The sun appears to track through twelve astrologically significant constellations (Aries, Taurus, Gemini, Cancer, Leo, Virgo, Libra, Scorpio, Sagittarius, Capricorn, Aquarius, and Pisces) each year, and to spend time each year with one of these constellations in the background. The zodiacal sign one is born under is determined by the background constellation the sun is in when one is born. Each sign allegedly signifies different personality characteristics. For example, Libra, the Balance, purportedly indicates cooperation and harmony. A person born when the sun is in Libra should, theoretically, be inclined to show a disposition to exhibit these traits.

The signs of the zodiac can be grouped in various ways. Three traditional groupings of the signs are: (i) the polarities – the signs are divided alternately into positive and negative; (ii) the quadruplicities – the signs are divided into cardinal, fixed, and mutable; (iii) the triplicities – the signs are divided into the categories of air, fire, water, and earth. These groupings classify the signs according to the various personality characteristics that are associated with each sign.

In addition to the sun's apparent annual motion around the earth, there is the daily rotation of the earth around its own axis. Just as astrologers

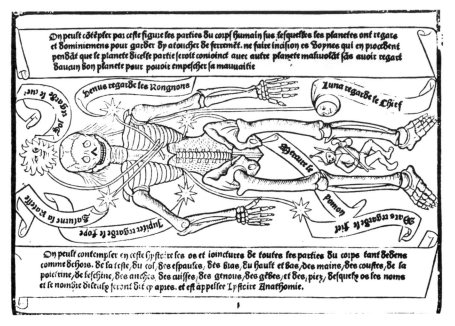

Fig. 13.1. Planetary skeleton. The planets were thought to control the major centres of the body, for example Venus ruled the kidneys and Jupiter ruled the liver. Woodcut from *Icy est le compost* (1483).

divide the sky into twelve regions (the signs of the zodiac), so do they divide the sky into twelve sectors called houses, for the daily rotation. The sun and planets travel through these sectors on a daily basis. The houses are not given names but are designated by numbers. Each house is associated with a particular aspect of life. The second house, for example, is associated with money and possessions. Of particular significance is the 'ascendant', the zodiacal sign that is rising in the east when an individual is born.

The planets are also associated with specific relationships with personality traits. For example, Mars is associated with combativeness and decisiveness. Each planet purportedly 'rules' one or more of the signs. Whenever a planet is in the sign that it rules it has a strong influence on the individual born at that time. Pluto and Mars, for example, are the rulers of Scorpio. Angular relationships (aspects) between the sun and planets are also considered to be significant. Sextiles (60° angles) and trines (120° angles) denote harmonious personality functioning, whereas squares (90° arcs) and oppositions (180° arcs) represent inharmonious personality functioning.

The interpretation of the horoscope is governed by a consideration of the following elements: the particular locations of the planets in the signs and houses, the position of the signs with respect to the houses, and by certain angles determined by the planets' positions relative to each other (Kelly, 1982).

Astrology is not regarded favorably by most scientists and scholars for several reasons. Some of these objections are as follows.

First of all, the scientific evidence is against most astrological claims. For example, zodiac signs, houses, planetary aspects, and rulerships are not supported by the evidence (Startup, 1985). Where the evidence does seem to offer support for astrological claims, as for example, Gauquelin's research on planetary positions and occupation and temperament, the relationships are weak and controversial (Eysenck & Nias, 1982; Kelly, 1982).

Second, we have been provided with no mechanism that can explain purported astrological relationships. Without such a mechanism, astrology clearly lacks in explanatory power, and also lacks the power to suggest what needs changing when fit with the evidence becomes a problem (Krips, 1982).

Third, results from physics and astronomy conflict with many of the claims of astrology. For example, Culver and Ianna (1984) point out that from the astrological literature itself the characteristics of the purported astrological planetary influences include the following:

(i) The influences bear no correlation to any intrinsic physical property of a given planet, including mass, size, density, rotation rate, surface and atmospheric composition, yet they are uniquely different for the sun, moon, and each of the planets. For example, both Jupiter and Pluto are claimed to equally affect us.

(ii) The influences are independent of line of sight planetary distances. Mercury is believed to still have effects on us even when it is on the opposite side of the sun.

(iii) Astrological space possesses a set of astrological influences of its own. Moreover, this space can be imprinted with crucial astrological points such as the midheaven and the ascendant.

We may quickly deduce from the above that the descriptions of the behavior of the astrological planetary influences as set forth by the astrologers themselves are totally at odds with the *empirically* determined characteristics of the four scientifically recognized types of interaction in the universe, that is, gravity, electromagnetism, weak and strong interactions.

In response to attacks by philosophers, scientists, and other scholars, astrologers have attempted to defend astrology by various arguments. It is our contention that the *a priori* arguments advanced in favor of astrology fail and that the empirical evidence presented on behalf of astrology, even if valid, does not support astrology as the majority of astrologers conceive it or as the general public conceives it. Our presentation can be considered a contribution to those who debate with astrologers: a set of cogent responses to common arguments advanced by astrologers in such debates.

1 Astrologers contend that because of its durability down the centuries, astrology must have 'a core of truth' to it

First of all, durability implies neither truth nor desirability in the human experience. Superstition, war, murder, slavery, and rape can all be included in the list of humanity's age-old endeavors, yet none of these has 'a core of truth', other than that they represent examples of the human spirit running amok. The world's major religions, Buddhism, Hinduism, Judaism, Christianity, and Islam have been most durable through long periods of human history, yet the 'truth' of each is substantially different from that of the others. The contention that the 'truth' of astrology arises from its durability is thus most suspect, and we must search elsewhere for an explanation of astrology's durability.

2 All, or most, cultures have developed a form of astrology, therefore there must be 'something to astrology'

Such a conclusion is not warranted. Even if it is true that astrology in some form is found in many cultures it does not follow that any of the tenets of astrology are true. There can be almost unanimous belief in something that is false, as early belief in the flat earth attests. Also, as Sagan (1980, p.51) points out, the false belief that the earth was at the center of the universe

> is the most natural idea in the world. The earth seems steady, solid, immobile, while we can see the heavenly bodies rising and setting each day. Every culture has leaped to the geocentric (earth-centered) hypothesis.

In addition, the forms of astrology believed in by ancients and modern astrologers are greatly varied. In Chinese astrology, the pole star was important and utilized twenty-eight constellations which bear no relationship to the constellations of interest in the Western zodiac. Chinese horoscopes ignored the ascendant (the zodiac sign that is rising in the east, when one is born), an event considered very important in Western astrology (Culver & Ianna, 1984). Also, as Eysenck and Nias (1982, pp. 33–4) point out:

> Because of precession there are in effect two zodiacs, the 'tropical' zodiac which is tied to the vernal point, and the 'sidereal' zodiac which is tied to the stars. The first is favoured by astrologers in the West, the latter by astrologers in the East. In Ptolemy's time the two zodiacs

coincided, but due to precession they are now nearly one sign out of step. However, the meanings have not changed, and signs of the same name still have the same meaning in both zodiacs. This means that almost opposite meanings can be given to the same piece of sky. Eastern astrologers also use other methods which are in conflict with Western methods; for example, most of them ignore the three modern planets now judged to be so essential in the West. ... [this] means essentially that if Western astrologers are right in making any particular interpretation, Eastern astrologers are wrong, and vice versa. Yet both sides claim to be extremely successful!

There tends to be general agreement on the symbolism of the planets (but not the number of planets that should be used!) and planetary angularity, however, this is *not* the case with most other central astrological tenets. For example, in respect to 'houses' there is 'disagreement on everything including number, sequence, position of peak strength, method of division, interpretation and validity' (Dean & Mather, 1977, p. 165).

Although the belief in some form(s) of astrology was (and still is) prevalent in most cultures, there is no single astrology that most people believe in. How could the fact that some people believe in one form of astrology justify the view that there must be 'something to astrology' when it is not clear what the valid tenets (if any) are?

3 Many great scientists of the past were astrologers which suggests that there is something to it

Many great scholars of the past and present are or were followers of all sorts of strange doctrines: Newton was a follower of alchemy; Alfred Russel Wallace and Conan Doyle were advocates of Spiritualism; Charles Dickens believed in phrenology, and so on. In addition, for every eminent scholar who believed in such a doctrine one can find another who did not. Whereas Kepler followed astrology, Galileo was a skeptic.

One can also ask *which* school of astrology did the scientist believe in? Ptolemy, the noted Greek astronomer thought that the influence of the heavenly bodies depended on their positions in the signs and houses and believed in natal astrology (astrology which uses the positions of the heavenly bodies to divine a person's character, likes and dislikes, etc.). Kepler, on the other hand, emphasized that astrology deals with *physical* effects and therefore abandoned the concept of houses and used constellations instead. He did not believe in natal astrology, he believed in what one might call 'social' astrology since he was interested in determining correlations between planetary positions and *mass* events (plagues, social upheavals, wars, etc.). So which astrological authority does one believe?

As the argument stands, it is just an appeal to authority. One needs to examine the evidence that the authority used to come to his conclusion and ask whether such evidence is methodologically sound and free of alternative explanations.

4 Astrology is an empirical science based on centuries of observation by ancient scholars

Most modern scientific methods consist of several components including experimentation, recognition of laws or behavior patterns in the experimental results, organizing the entire collection of laws into an overall theory, and the predictive testing of that theory. Upon investigation of the astrologers' conduct of their 'science', we find wholesale violations of this

Fig. 13.2. Title page to the *Astrologischer Spiegel* by Johann Georg Sambach (Nuremberg, 1680), showing how the signs of the zodiac affect the body according to the time of birth.

method (Culver and Ianna, 1984). For example, rather than weeding out those astrological views which are incompatible with empirical results, astrologers have instead sought to include virtually every possible factor in the horoscope in a fashion not terribly unlike the medieval astronomer adding another epicycle to a planetary motion when the discrepancy between observation and theory became painfully manifest. As a result, instead of a single astrological system which is tailored to best fit empirical results, we are instead treated to a multiplicity of different astrologies having a multiplicity of conflicting central tenets. After two thousand or so years one would expect more agreement! In fact, a number of astrological texts will state that the best astrological system to use is the one with which you feel most comfortable! It was the rejection of such an approach in favor of the empirical method that provided the central theme of the Scientific Revolution.

There are also serious historical questions concerning the 'empiricism' of astrology. As a simple example, when Ptolemy wrote the *Tetrabiblos*, the world's population was about 50 million people. An astrologer of the day with access to the horoscope of *every* person on the globe at that time could conduct a meaningful statistical test of the distribution of no more than six planets by astrological sign at birth. Obviously no tests could possibly have been made of additional astrological factors such as the houses, aspects, and other planets (Culver & Ianna 1984, Ch. 10).

In a somewhat similar vein, the planets Uranus, Neptune, and Pluto are all regarded as most important factors in most horoscopes, yet for centuries no astrologer was able to come forward and deduce the existence and position of any of these planets on the basis of horoscopic astrology. One particularly interesting response is the claim made by some astrologers that the astrological influences of a given planet do not begin until the first moment of that planet's discovery! (e.g. Goodman, 1968).

5 A variety of extraterrestrial influences on the earth and organisms has been documented, lending general support to astrological theory

Astrologers point out that studies exist that support a relationship between biological effects of weather and climate on living organisms, the relationship between the moon and earthquakes, geomagnetic field effects on the behavior of animals, and the effects of atmospheric electrical factors. In addition, the influence of sunspots and the solar wind with possible interplay with the planets (e.g. Phillips, *et al.*, 1980; Sulman, 1980) is considered evidence for astrology by astrologers. However, apart from the fact that such claims are still controversial in that the reliability and strength of such effects is very debatable, it is a far jump from such claims to the thesis that the planetary configurations at one's birth and at any particular time reflect

Fig. 13.3. An astrologer. One of a collection of seventeenth-century French prints showing different craftsmen and their tools, skills or attributes.

current events that occur with individual human beings that astrologers deal with. A number of other similar relationships have been reported (Culver and Ianna, 1984, pp. 197–98), but in each case, the possible 'celestial influences' bear absolutely no resemblance whatsoever to the astrologer's descriptions and predictions. In fact, if any conclusion is to be drawn concerning astrology from such studies, it is that *all* of the traditional horoscopic systems should be discarded.

6 There are many empirical studies that provide support for specific astrological tenets

In any field of science there are studies that produce results that are anomalous. Some of these studies may later contribute significantly to science; others will later have been shown to be spurious. An example of the latter is polywater. Anomalous water, or polywater, was at first thought to be a variety of H_2O with a different molecular structure than ordinary water which gave it properties said to be quite different from ordinary water. Although, at first, there were successful replications of the original finding, it later became clear that polywater was not a new form of water, but rather, water contaminated with various impurities (Franks, 1981). The example of polywater underscores the importance of not placing very much faith in a single experiment or a series of experiments but rather in successful replications by *independent* investigators around the world. In addition, the critical examination of the experimental protocols by competent investigators is needed to rule out plausible alternative explanations of the results before the anomaly can be accepted as such.

The field of astrology is full of claimed 'startling evidence' for astrology that was later shown to have been based on inappropriate statistical or methodological procedures or susceptible to 'normal' alternative explanations (Eysenck & Nias, 1982).

More recently, a series of studies supporting astrological claims have been shown to be methodologically or statistically defective. In 1978, widespread media attention was given to a study published in a leading psychological journal (Mayo, White & Eysenck, 1978) that claimed to have found a (weak) relationship between zodiac sign and various personality characteristics. Subsequent research, which involved one of the researchers (Eysenck), found that the results were due to self-attribution, that is, those who know the meaning of their zodiac sign (and other astrological tenets) measurably shift their self-image and perhaps behavior toward the descriptions given by astrology, whereas those who are ignorant of the meaning display no measurable effect (Eysenck & Nias, 1982). The work of Nelson claimed that planetary positions correlate with the quality of short wave radio propagation and has been extensively quoted by astrologers. It has now been shown to have been based on statistical artifact, and, in fact, does not work (Dean, 1983). Similarly, astrologers have appealed to studies that suggest that moon phase is linked to various human behaviors such as homicide rate, drug overdose and number of automobile accidents. Several in-depth reviews (Rotton & Kelly, 1985; Kelly, Rotton & Culver, 1986) have shown there is no good evidence for a physical link between human behavior and phases of the moon. Finally, in a four-part series in March, 1984, the prestigious British newspaper, the *Guardian*, presented the results of a massive study of the relationship between occupation and sun sign. Some astrological effects were claimed to have been shown. A reanalysis showed that the results could be explained by statistical fluctuations and self-attribution (Dean, Kelly, Rotton, Saklofske, 1985).

There is one notable exception. The research of Michel Gauquelin (1983) seems to have stood up quite well to criticism and replication. Although Gauquelin has found *no* evidence for any system of zodiac signs, planetary aspects or any evidence of the ability of the horoscope as a whole to be of any use in predicting human behavior, he claims to have found a relationship between some personality characteristics and the diurnal positions of the inner planets (but *not* the outer ones – Uranus, Neptune and Pluto). Gauquelin's work is not free of problems however. First of all, there is, so far, no satisfactory mechanism to explain his findings (Krips, H., 1979, 1982). Second, not all research has supported his theories; although it must be acknowledged that the empirical evidence, on balance, is in favor of Gauquelin's position. An example of negative evidence concerns Dean's study (1981a; see also, 1985a, b) involving subjects with extreme extrovert and introvert scores on the Eysenck Personality Inventory. There was no evidence that Gauquelin's theory could predict these personality dimensions in this study. Clearly, more research by independent investigators on new samples of subjects needs to be conducted to clarify the findings in this particular area. A final point regarding Gauquelin's research is that his results are so sufficiently different from, and weaker than, what has traditionally been held by astrologers that they can hardly provide a basis for popular belief (Startup, 1985).

7 Astrology works

No claim is more commonplace in the astrological literature than the simply stated 'Astrology works'. Astrology books (e.g. Parker and Parker, 1984) usually contain case studies of horoscopes of famous individuals (e.g. Oscar Wilde, Jimi Hendrix, John Kennedy, etc.) purportedly showing how well astrology described their lives and character. Despite its powerful simplicity, close examination of this oft-stated canard strongly suggests otherwise. In a study made in 1979 of over three thousand predictions made by astrologers, barely ten percent were fulfilled even when predictions which could be attributed to shrewd guesses, vague wording, or 'inside' information were counted as predictive successes (Culver & Ianna, 1984, 169–70).

One of the central claims of most astrologers is that one cannot use various astrological factors (signs, houses, aspects, etc.) singly, one has to test astrology with the whole horoscope. As Dobyns (1974, p. 131; see also Niehenke,

Fig. 13.4. A talisman for love. This is said to be wonderfully useful in obtaining success in love adventures. It must be made in the day and hour
of Venus, when she is favourable to the planet Mars. It should be
made in pure silver, or purified copper. If Venus is in the sign of
Taurus or Libra (her particular houses) this is even better. From
The Astrologer of the Nineteenth Century (London, 1825).

1983) tells us: '... the one primary rule in astrology is that no factor can be
taken out of context without a real danger of losing the meaningful gestalt'.
However, this claim does not stand up to either theoretical argument or
empirical evidence.

Theoretically the claim is dubious. As Geoffrey Dean (1979, p. 93) has
pointed out:

> ... In traditional astrology the minimum package of 10 planets (in-
> cluding the moon) in 12 signs, 12 houses, and making 5 major aspects,
> provides nearly 500 different factors of which about 30–40 are also
> present in the average horoscope. Hence to claim, as many astrologers
> do, that tradition is the result of millenia of empirical observation, is
> to claim that the meaning of each of 500 factors can be deduced when
> any 40 can be present at the same time. This is clearly untenable.

More importantly, the evidence indicates that horoscopes do not work in
the way that astrologers contend they do. First of all, astrology has an almost

unlimited scope to describe anything in retrospective over a wide range of applications. For example, Dean and Mather (1977, p. 25) tell us:

> [The astrologer] Henderson presents a detailed astroanalysis of the character of Lenin and the events in his life. The birth data quoted are correct but the chart used corresponds to a date 12 days earlier ... as a result about half the significators are wrong yet they match the character and events perfectly.

Similarly, at a conference dealing with research into astrology in Britain, Dean (1981) showed that British singer Petula Clark's horoscope matched her biographical details exactly; the horoscope was then revealed to be that of U.S. murderer Charles Manson. Similarly, Niehenke (1983), recently came across three different astrological publications, each containing a different horoscope of Beatle John Lennon, based on *three different birth times*, all indicating 'definitely' Lennon's sudden death. Dean and Mather (1977, pp. 19, 28–31, 211; see also, Culver & Ianna, 1984, Ch. 7) provide a list of further examples from a variety of horoscope applications, including character analysis, correlation with events, political horoscopes, and so on. It is quite clear that astrologer's appeals to personal successes cannot carry very much weight.

Second, investigations of the validity of personality interpretations based on horoscopes have found that people were unable to distinguish between authentic interpretations and false interpretations. In other words, wrong horoscopes are accepted as readily as right horoscopes. In a review of studies examining individual's ratings of two or more horoscopes for accuracy, of which one was theirs, in each study people were unable to pick the authentic interpretation better than would have occurred by chance (Dean, 1986). It should be acknowledged at this point that the research of Geoffrey Dean has been of outstanding importance in providing concrete empirical data on astrology. In a more rigorous study, Dean (1986) altered individual's horoscopes to make them as opposite in meaning to the authentic horoscopes as possible: 'Thus extraverted indications were substituted for introverted, stable for unstable, tough for tender, ability for inability, and so on.' (p. 3.) It was found that reversed interpretations were accepted just as readily as authentic interpretations. Dean (1986, p. 2) points out that these results cannot be explained by the subjects not knowing themselves, because when the same test is applied to personality inventories, the correct profile tends to be chosen (Green *et al.*, 1978, Hampson *et al.*, 1979).

Tyson (1984) expanded his experiment by getting subjects *and* someone who knew them well, to try to identify which one of a number of cast horoscopes belonged to the subject. Both the subjects and the nominees were unable to do this. Ianna and Tolbert (1985) tested the ability of a prominent

Fig. 13.5. The planets, signs of the zodiac and the departments of life attributed to each house division. Sixteenth-century woodcut.

astrologer to pick the correct horoscope out of four on the basis of studying the face and build of people. In spite of the astrologer's claims that he could do so, a carefully controlled test with 28 students yielded chance results.

Third, tests of the ability of astrologers using the complete horoscope to describe character have not been particularly successful. Clark (1970) initiated such tests by examining the abilities of astrologers to match individual horoscopes and various qualities of the respective individuals such as career, case history, etc. He carried out three such tests, and each gave positive results in favor of astrology. Gauquelin (1983), however, finds Clark's studies unconvincing. He writes:

> Frankly, I don't think Clark's test proves anything. For instance, he refers to a successful experiment with 50 astrologers from England, the U.S.A. and elsewhere without giving their names, thereby making it impossible to measure their skill again in a second test along the same lines. Moreover, the French astrologers who tried Clark's test – all serious professionals – failed it completely ... The failure of the French astrologists remains inexplicable: there is nothing to show that their qualities as practitioners were in any way inferior to those of Vernon Clark's colleagues (pp. 137–139).

Since Clark's study was published, other similar tests have been conducted, with mixed results (Dean & Mather, 1977, pp. 544–51). However, most of these tests have involved methodological flaws such as failing to control for alternative explanations such as self-attribution. In addition, the studies have too small a number of horoscopes (Dean, 1985*b*; Eysenck & Nias, 1982).

In a massive, recent study that gets around these objections, Dean (1985, *a, b*) obtained no evidence for astrology. He tested whether astrology could predict two of the most important personality factors, namely, extraversion and emotionality in ordinary people. He tells us:

> Extraversion and emotionality should be among the easiest of things to see in a chart (horoscope), because as the four temperaments they have been around for at least two thousand years. And they are far less open to external influences than say business success or a miserable childhood ... if astrologers cannot see extraversion and neuroticism in a chart, there would not seem to be much hope for anything else. (p. 2)

To this end, subjects with extreme scores on the *Eysenck Personality Inventory* were selected from over one thousand people. The average pair of opposite extremes used in the test was roughly equivalent to the two most extreme persons in a random sample of 15 adults. As Dean (1985*a*, p. 4) points out: 'If astrology cannot correctly judge the two most extreme persons in every 15, there is clearly no hope for the other 13.'

In the first experiment, Dean (1985*a*) tested whether computer analyses of astrological factors such as tropical signs, decans, elements, sidereal signs, aspects, harmonies, Gauquelin plus zones, and angularity, both individually and in combination could predict extraversion and neuroticism (emotionality) in extreme subjects. None of the astrological factors, either singly or in combination performed better than chance. In a second experiment, Dean (1985*b*) tested whether 45 astrologers using the whole chart could predict extraversion and emotionality in ordinary people. He found that the astrologers did no better than chance. In addition, there was no effect of the astrologer's experience, sex, personality or technique.

Fourth, and most importantly, Dean found there is poor agreement among astrologers regarding what astrological factors should predict various personality characteristics and even astrologers using the same technique (the same factors) show little agreement – this suggests that each astrologer's technique and interpretation are highly individual. In other words, the results of Dean's study showed that astrologers do not usually agree on what a chart (horoscope) indicates, even when using the same factors!

Carlson (1985) checked astrologer claims that they can tell from natal charts what people are really like and how they will fare in life. Carlson asked astrologers to interpret natal charts for 116 unseen 'clients'. Carlson's research involved 30 American and European astrologers considered by their peers to be among the best practitioners of their art.

For each client's chart, astrologers were provided three anonymous personality profiles – one from the client and two others chosen at random – and asked to choose the one that best matched the natal chart. All personality profiles came from real people and were compiled using questionnaires known as the California Personality Inventory (CPI). Despite astrologers' claims, Carlson found those in the study could correctly match only one of every three natal charts with the proper personality profile – the very proportion predicted by chance.

8 Scientists and other scholars are not familiar with astrological theory and practice and therefore are not qualified to evaluate astrology

According to this argument only astrologers are qualified to discuss and evaluate astrological practice. However, the argument that one has to have experienced or have an internal familiarity with a phenomenon or theory or practice in order to be able to write or think intelligently about it is absurd. It would follow that someone who has never murdered anyone is not qualified to evaluate or examine the case of someone who has.

Moreover, astrology is concerned with human personality and affairs and the individual's prospects as the future approaches. The several disciplines concerned with human beings (e.g. psychology, sociology, biology) have the same concerns. If astrology had results of substance, these disciplines would be very interested in learning what they are.

Fig. 13.6. Astrologers casting a horoscope for the child being delivered in the foreground. From Jacob Rueff's *De conceptu et generatione hominis* (Frankfurt-am-Main, 1587), E.P. Goldschmidt and Co. Ltd.

A related argument is that one has to actually work with horoscopes before one can observe the powers of astrology (Dean, M. 1980, Dijkstra, 1983; Niehenke, 1983, p. 235). In other words, while astrological language constitutes a public manifestation of astrology, one does not *understand* what astrology really means unless one actually participates in the activity of constructing and interpreting horoscopes for clients. This claim would seem to be false; there are those (e.g. Michael Gauquelin, Geoffrey Dean) who learned the language of astrology and came to understand the practice of astrology without becoming believers in the validity of horoscopes and much astrological practice. The argument is further weakened when one considers, as was pointed out earlier, that astrologers show poor agreement (Dean, 1985*b*). One could also find those who after, perhaps, studying more science or philosophy, came to lose their belief in astrology. It would be rather absurd to say that when they lose their belief in astrology they also lose their understanding of astrology. This argument seems to be a move of desperation to try to preserve their position from critical scrutiny. (See Nielsen, 1983.)

9 Astrology is not a science but an art or a philosophy

As Mather (1979, p. 106) pointed out:

> The initial assumptions of scientists and astrologers were not so very different until very recently. Some astrologers however, fearing that science was catching up with them, have backtracked very rapidly, creating a smokescreen of symbolism, inner reality, holistic understanding, etc.

As a consequence, the majority of astrologers reject a scientific approach to astrology in favor of symbolism and holistic understanding (Dean & Mather, 1977, p. 2). Some astrologers hold that the operative principle governing the universe, or at any rate the parts of it astrology purports to study and disclose, is what is called synchronicity. This idea is seldom stated with full clarity, but it is evidently the notion that patterns in one part of the universe are mirrored, echoed, or paralleled in others, acausally. Though there is (the theory continues) no causal relationship between one pattern and its parallel(s), one can learn about the one from the other. As it happens, these astrologers hold, patterns of human affairs are paralleled in the stars, each successive astrological time period on earth reflecting and being illuminated by appropriate stellar configurations. However, the notion of synchronicity is very problematic. Causal relationships need not be as simple as the classic case of one billiard ball striking another. Two patterns may have a common cause, though neither is the cause of the other. Day and night are examples of this. If one pattern is in any sense significantly or interestingly replicated in another pattern, this will invariably invite scientific scrutiny. *Why* is the one phenonomenon 'reflected' in the other and how do the two chains of events keep their perfect symmetry? Even if one has not caused the other, the working assumption of inquiry will, naturally, and unyieldingly, be that there is *some reason*, some ground of connection between the two, and that the mechanism of connection must in principle be discoverable. The concept of synchronicity therefore seems totally unhelpful in describing or explaining what studying the heavenly configurations, in the astrological mode, may say about human life.

In addition, it is difficult to provide a formulation of 'holism' that is not either vacuous, false, or impossible to implement in practice. If by holism we mean that our character and life events require a consideration of all contributing factors – those which are psychological, social, physical, emotional and perhaps spiritual, this is not a new revelation. Insofar as multiple factors are known, physicians and psychologists routinely address such issues (often by working together) (see Gylmour & Stalker, 1983 on this topic). In the past in psychology there was an overemphasis on reductionism and a tendency to only seriously consider the results of experimental studies. However, over the last decade or so there has been an increased interest in naturalistic studies and a consideration of the development of the human being in his whole social, developmental and historical perspective (e.g. Elder & Rockwell, 1979).

One may also mean by holism the view that everything in the universe or solar system is related to human activity. Since we cannot attend to everything in the solar system, this interpretation would seem to be bankrupt in practice. Kollerstrom (quoted by Schneider, 1982, p. 34) contends that astrology is an art that evaluates how the whole solar system in the zodiac at a particular time is related to the whole person at that moment. These astrologers deny that an analytical and quantitative approach to verification is helpful. Taken literally, Kollerstrom's view says very little. Everything has (logically, *must* have) *some* relation to every other existing thing, so everything in the solar system, and the solar system as a whole, will have some relation to every human being. But which relation or relations? Some will be wholly trivial (e.g. numerical diversity). It is presumably a significant causal relationship, a relationship of influence or parallelism, that will be wanted. But how can a relationship of this sort fail to fall under the purview of analytical empirical investigation? In addition, does 'everything' include all the moons of all the planets, the asteroids and any comets that happen to be in our solar system at the time? Such a doctrine does not distinguish relevant from irrelevant factors, important from unimportant relationships.

The claim that astrology is an art or craft is also problematic. Flying an airplane or teaching are often considered arts in the sense that it is difficult or impossible to advance a set of directives or principles explaining how to do these things. However, there *are* criteria by which one can evaluate flying an aircraft or teaching. On the other hand, how does one evaluate the art or craft of astrology? By the aesthetic appearance of the horoscope or its length or by client satisfaction? If the latter is sufficient, then the evidence suggests that astrologers do not have to obtain accurate birth data or labor over a horoscope to write a satisfying astrological reading. They can satisfy the

Fig. 13.7. Astrology has attracted artists of all kinds. Two scenes from the ballet 'Horoscope', first produced in 1938, which portrays a man and a woman born under two conflicting signs and brought together by the moon in their mutual sign Gemini.

client by the use of general statements that are vague or ambiguous (e.g. 'You are susceptible to flattery and the influence of others. You are inclined toward nervousness and are over-sensitive'.), and favorable (e.g. 'You are an independent thinker. You have a great deal of unused capacity.') In addition, the astrologer can 'flesh out' the reading by observing the client's body language: how he/she is dressed and groomed and what the client's posture and movements reveal (Dickson & Kelly, 1985; Dean & Mather, 1977; Hyman, 1981).

Another ploy is to describe astrology as *not* a science but a philosophy or alternative world-view with its own rules and procedures. Shallis (1981) defends this view. He contends that astrology springs from a religious world view and is just another way of looking at the world. Shallis argues that astrological principles are intuited rather than based on observation and reason. But even if 'intuited', one still needs some *reason* to believe something. Why should anyone believe a given astrological claim, or set of claims? No doubt, everyone has a right to believe whatever they want to believe. Nonetheless, most will want to believe on the basis of *some reasons*. To merely claim that astrological principles are based on intuition will not do – intuitions notoriously differ and conflict (we have a plethora of conflicting systems of astrology!) How do we settle these differences – by appealing to other intuitions?

A related approach is adopted by Alexander (1983) who contends that astrology is like poetry rather than science, and that interpreting a horoscope is more like writing a poem than doing an experiment. He tells us that the meanings of chart factors are more like metaphors or similes than like testable claims. The analogy with poetry is suspect. Even though poetry is often expressed in complex language, it is very often a poet's intention to convey some truth about human nature or an important issue (e.g. an anti-war poem). So poetry does frequently make a knowledge claim, albeit in an indirect way. Although the poet does not do the verification, the reader can. For example, if a poem conveys the view that all humans are basically dishonest or have an inextinguishable propensity for war we can evaluate these claims in an empirical way. Even if these claims are suspect the poem is not refuted since the claims are only part of what the poem is intended to do. So, often, poems do contain assertive, cognitive components.

In 1984, a group of astrologers in California put forward the following definition of astrology:

> Astrology is the *philosophy* that postulates a relationship between relevant celestial phenomena and/or processes and certain terrestrial affairs (Editor, 1984, p. 31, Italics ours).

While this approach is an improvement on previous ones, it does not, as no doubt many astrologers would wish, make astrology immune from criticism. We can still ask of a philosophy: How clear and consistent are its tenets? How compatible are its claims with the knowledge acquired in other related fields? Does it have an empirical foundation? Most philosophies still make public-ally testable (empirical) claims. For example, most religious philosophies make historical claims and many make other claims about faith healing, etc. that can be approached in an empirical way. In the absence of testable, empirical verified claims astrology becomes bankrupt in practice.

There is a further issue here: the vast majority of clients of astrologers do not view astrology in ways described by Shallis or Alexander. If one examines how astrology has been used over the centuries one obtains an overwhelming human interest in a concern to know the future and a concern with present affairs, 'Should I marry X? Should I make this trip? What are my potentials?' and so on. Such concerns, to predict and describe, are precisely those of modern science!

Conclusion

The past poor quality of experiments offered as proof of astrological claims should make the scientific community very wary of newly presented discoveries that support astrology. On the other hand, it would be unscien-tific to dismiss such claims *a priori*. Such claims, however, should be rigor-ously examined according to two fundamental criteria: first, such experi-ments should be well-designed to eliminate alternative non-astrological explanations and should be replicable by independent scholars on new populations. It would also be preferable if the scholars who do the repli-cations are either neutral or skeptical of astrological claims. It should also be the case that advocates of astrology work *with* such scholars in designing experiments to replicate studies claiming positive results (see Eysenck, 1982). Second, successful replications should demonstrate a *strong relationship* between astrological factors and human affairs. As Hays (1981, pp. 293–4) points out:

> ... People pay too much attention to [whether there is evidence of a real effect] and too little to the degree of association the finding represents. This clutters up the literature with findings that are often not worth pursuing, and which serve only to obscure the really im-portant predictive relations that occasionally appear. The serious scientist should ask not only 'Is there any association between X and Y?' but also 'How much does my finding suggest about the power to predict Y from X?' Much too much emphasis is paid to the former, at the expense of the latter question.

The discovery of a weak astrological relationship would not have any practical value (although it may have theoretical value) and would not support astrology as it is practised. Clients consult astrologers on the *assumption* that there is a strong relationship between celestial events and their affairs. If, as the majority of scientists contend, most, if not all, of our behavior is due to non-astrological factors such as our heredity, physical and cultural environment, including our upbringing, significant social relationships, chance events and encounters, and our own decisions, then a weak astrological effect would contribute very little to self-understanding or our understanding of others.

If astrology has little validity, it is legitimate to ask why astrology has such a large following in both the east and the west. First of all, this is the age of science, and astrology has the superficial appearance of a science. The terminology of science (quark, black hole) sounds no more (or less) mysterious than the terminology of the astrologers. In addition, astrology uses mathematics, as do most sciences, and the construction of a horoscope looks as precise as the blueprint of a new building. It is understandable that people believe that astrology is a science.

Second, most newspapers and many magazines contain daily or monthly horoscopes. Many people, no doubt, believe that the media would not include horoscopes unless they had some validity.

Third, those who have been to astrologers are often convinced that astrology works. However, acceptance of horoscopes depends on situational and psychological factors rather than on the validity of the horoscope. For example, horoscopes contain Barnum statements, these vague and ambiguous statements that tend to be accepted by people as descriptive of their unique personalities (Dickson & Kelly, 1985). Astrologers increase horoscope acceptance even more when they emphasize socially desirable characteristics in the client, base the horoscope on data as precise as possible (e.g. hour of birth rather than just day of birth) and utilize information obtained from observation of the client's appearance and behavior to advantage (Dean, 1981b). Finally, being well-groomed, using some incomprehensible terminology, talking fast, and appearing confident and sure of the information one is imparting, will guarantee success with the vast majority of clients (Kelly & Renihan, 1984). Ironically, it appears that if an astrologer wishes to become more successful, he should study less astrology and more psychology and sociology! Fourth, cognitive processes may play a role in the acquiring and maintenance of belief in astrology. Glick and Snyder (1986) found that believers in astrology were more likely than skeptics to distort facts that did not support astrology and exaggerate any positive evidence for astrology they received. Fifth, and of no small significance in its own right, astrology has been for a very long time a profession and practice of a considerable number of men and women who derive their incomes from people wishing astrological service. That is, we ought to look at the phenomenon of astrology not just from the perspective of the consumer (of astrology), but also from that of the producer. Astrologers can and some do make handsome livings – in some cases immense wealth – from the willingness of members of the public to part with their money for what they believe astrology will or might do for them. There is therefore considerable motivation for professional astrologers to display (which may in fact mean to cloak) their theories, and their practice, in a light most calculated to win attention, and if possible, credence, both from the lay person, and a sympathetic theorist.

There are other reasons for the continuing wide influence and acceptance of astrology. Astrology was already a highly developed body of theory and speculation when the great age of science began in the seventeenth century. Indeed, astrology and astronomy arrived hand in hand in the first generations of truly empirical scientific advance. Kepler, for example, made most of his income as an astrologer. The natural science of Tycho Brahe, Kepler, Galileo, Boyle, and Newton had virtually nothing to say about human beings and their concerns; and what we know as social science only began to develop in the nineteenth century. Where science was silent, astrology and similar approaches continued, and still continue, to fill the gap created by the widespread and longstanding human concern to know, and if possible, control, human destiny, both collective and individual. The very remoteness of much scientific development from familiar and paramount human concerns may, as well, have failed to encourage a deeper and more diffuse acquisition of the *empirical* bent, which is the core of science. Until the claims of astrologers can be repeatedly verified the stars should be used for the navigation of ships and not lives.

References

Alexander, R. 1983. Letter to the editor. *Correlation*, **3**, 34–5.

Carlson, S. 1985. Double-blind test of astrology. *Nature*, **318**, (Dec. 5), 419–25.

Clark, V. 1961. Experimental astrology. *Aquarian Agent*, **1**, 9, 22–3.

Culver, R., & Ianna, P.A. 1984. *The Gemini Syndrome*. Buffalo, N.Y.: Prometheus Books.

Dean, M. 1980. *The Astrology Game*. Don Mills, Ontario: Nelson Foster & Scott.

Dean, G.A. 1981 (*a*). Planets and personality extremes. *Correlation*, **1**, 2, 15–18.

1981 (*b*). The acceptance of astrological chart interpretations: a simple test of personal validation using reversed charts. Paper presented at the 2nd Astrological Research Conference, Institute of Psychiatry, London, England. November.

1983. Forecasting radio quality by the planets. *The Skeptical Inquirer*, **8** (Fall), 48–56.

1985 (*a*). Can astrology predict E(xtraversion) and N(euroticism)? Part 1. Individual factors. *Correlation*, **5**, 1, 3–17.

1985 (*b*). Can astrology predict E(xtraversion) and N(euroticism)? Part 2. The Whole Chart. *Correlation*, **5**, 2, 2–24.

1986. Does astrology need to be true?' *The Skeptical Inquirer*, **XI**, 2, 166–83.

Dean, G.A. & Mather, A. (Ed.). 1977. *Recent Advances in Natal Astrology*. Bromley, Kent: The Astrological Association.

Dean, G.A., Kelly, I.W., Rotton, J., & Saklofske, D.H. 1985. The Guardian astrology study: A critique and reanalysis. *The Skeptical Inquirer*, **IX**, 4, 327–338.

Dickson, D.H. & Kelly, I.W. 1985. The Barnum effect in personality assessment. *Psychological Reports*, **57**, 367–82.

Dijkstra, B. 1983. The practice of astrology. *Astro-Psychological Problems*, **1**, 3, 30–32.

Dobyns, Z. 1974. Review of Gauquelin 'Cosmic Influences on Human Behavior.' *Psychology Today*, **9**, 131.

Editor. 1984. Conference Report. *Astro-Psychological Problems*, **2**, 1, 30.

Elder, G. & Rockwell, R.C. 1979. The life-course and human development: an ecological perspective. *International Journal of Behavioral Development*, **2**, 1–21.

Eysenck, H.J. 1982. Methodology and the Vernon Clark experiment. *Astro-Psychological Problems*, **1**, 1, 27–29.

Eysenck, H.J. & Nias, D.K.B. 1982. *Astrology: Science or Superstition?* London: Maurice Temple Smith.

Franks, F. 1981. *Polywater*. Cambridge, Mass.: MIT Press.

Gauquelin, M. 1983. *Birthtimes*. New York, N.Y.: Hill and Wang.

Glick, P. & Snyder, M. 1986. Self-fulfilling prophecy: the psychology of belief in astrology. *The Humanist*, (May/June), 20–25 & 50.

Goodman, L. 1968. *Linda Goodman's Sun Signs*. New York, N.Y.: Bantam Books, Inc.

Greene, R.L., Harris, M.E. & Macon, R.S. 1979. Another look at personal validation. *Journal of Personality Assessment*, **43**, 419–423.

Gylmour, C. & Stalker, D. 1983. Engineers, cranks, physicians, magicians, *The New England Journal of Medicine*, **308**, 16, 960–964.

Hampson, S.E., Gilmour, E. & Harris, P.L. 1978. Accuracy in self-perception: the 'fallacy of personal validation.' *British Journal of Social and Clinical Psychology*. 17, 231–235.

Hays, W.L. 1981. *Statistics* (3rd ed.) New York, N.Y.: Holt, Rinehart & Winston.

Hyman, R. 1981. The psychic reading. In R.A. Sebeok & R. Rosenthal (Eds.) *The Clever Hans Phenomenon*. New York, N.Y.: Academy of Sciences. pp. 169–181.

Ianna, P.A. & Tolbert, C.R. 1985. A retest of astrologer John McCall. *The Skeptical Inquirer*, **IX**, 2, 167–170.

Kelly, I.W. 1982. 'Astrology, Cosmobiology, and Humanistic Astrology.' In P. Grim, (Ed.) *Philosophy of Science and the Occult*. Albany, N.Y.: State University of New York Press.

Kelly, I.W., Rotton, J. & Culver, R. 1986. 'The moon was full and nothing happened:

a review of studies on the moon and human behavior and lunar beliefs.' *The Skeptical Inquirer*, **X**, 2, 129–143.

Kelly, I.W. & Renihan, P. 1984. Elementary credibility for executives and upward mobiles. *The Canadian School Executive*. **3**, 10, 16–18.

Krips, H. 1982. Comments on Curry. *Zetetic Scholar*, **9**, (March), 63—64.

____ 1979. Astrology: Fad, Fiction, or Forecast? *Erkenntnis*, **14**, 373–392.

Mayo, J., White, O. & Eysenck. 1978. An empirical study of the relation between astrological factors and personality. *Journal of Social Psychology*, **105**, 229–236.

Mather, A. 1979. Response to critics. *Zetetic Scholar*, 1979, **1 & 2**, 15–18.

Nelson, J.H. 1951. Shortwave radio propagation correlation with planetary positions. *RCA Review* (March) 26–34.

Niehenke, P. 1983. The whole is more than the sum of its parts. *Astro-Psychological Problems*, **1**, 2, 33–37.

Nielsen, K. 1983. Skepticism and belief: a reply to Benoit Garceau. *Dialogue*, **XXII**, 391–403.

Parker, D. & Parker, J. 1984. *The New Compleat Astrologer*. New York, N.Y.: Harmony Books.

Phillips, R.D. *et al.* 1980. *Biological Effects of Extremely Low Frequency Electromagnetic Fields*. Oak Ridge, TN. Technical Information Center, Department of Energy.

Rotton, J. 1985. Astrological forecasts and the commodity market: random walks as a source of illusory correlation. *The Skeptical Inquirer*, **IX**, 4, 339–346.

Rotton, J. & Kelly, I.W. 1985. 'Much Ado About the Full Moon: A Meta-analysis of Lunar–Lunacy Research.' *Psychological Bulletin*, **97**, 2, 286–306.

Sagan, C. 1980. *Cosmos*. New York, N.Y.: Random House.

Schneider, M. 1982. Review of Eysenck & Nias 'Astrology: Science or Superstition?' *Astro-Psychological Problems*, **1**, 1, 34.

Shallis, M. 1981. The problem of astrological research. *Correlation*, **1**, 2, 41–46.

Startup, M. 1985. The astrological doctrine of 'aspects': a failure to validate with personality measures. *British Journal of Social Psychology*, **24**, 307–315.

Sulman, F.G. 1980. *The Effect of Air Ionization, Electric Fields, Atmospheric and Other Electric Phenomena on Man and Animals*. New York, N.Y.: C.C. Thomas Pub.

Tyson, G.A. 1984. An empirical test of the astrological theory of personality. *Personality and Individual Differences*, **5**, 247–250.

14

Astronomy and science fiction

ALLEN I. JANIS

Astronomy is a natural background for science fiction. What better setting for imaginative ideas than on other worlds or among the stars? Of course as more is learned about astronomy, its use in fiction changes – the moon is certainly not as popular a setting as it was before we learned how barren it really is, and even the other planets of our solar system no longer seem likely places to find interesting unearthly creatures. But the Universe is vast! If scientific discoveries have made once favorite locales less suitable for the stories, the author only has to move them farther away.

This vastness creates problems, however; at least it does for authors that don't want to fly in the face of scientific knowledge. Einstein taught us that it is impossible to travel faster than the speed of light in vacuum, a speed usually denoted by the letter c. But at speed c, it would take four years to travel to the nearest star and tens of thousands of years to reach the center of our galaxy. So if science-fiction authors want to have their characters flitting among the stars or, worse, among galaxies, they must come up with ways to avoid having these characters die of old age before anything interesting can take place.

One way to solve this problem comes from Einstein's own work. His theory of relativity shows that if you travel fast enough (but, of course, always less than c), you can get as far as you wish in as little of your own time as you wish. How can this be possible? Let us look at some simple relativity theory to see how it comes about. To keep things simple, let us suppose that gravitational effects are not important for our considerations. Then the version of relativity that applies is known as special relativity, and is founded on two basic postulates known as the principle of relativity and the constancy of the speed of light in vacuum.

The principle of relativity asserts that the fundamental laws of nature are the same for all unaccelerated observers. (It will be important to remember that 'no acceleration' means no change in either speed or direction of motion.) Suppose, for example, that you and I are both unaccelerated, but we are moving past one another with a relative speed of, say, $c/2$. (Never mind the practical question of how we could achieve a speed of half the speed of light! This discussion is, after all, in the context of science fiction.) That is, I measure your speed relative to me and you measure my speed relative to you, and we each find the other's speed to be $c/2$. It follows from the principle of relativity that there is nothing either of us could do to learn whether one of us is 'really' at rest and the other is 'really' moving at $c/2$. The only thing we can establish is that our relative speed is $c/2$. Either of us could consider ourself to be at rest, if that were convenient; the concept of absolute rest is meaningless.

The postulate of the constancy of the speed of light in vacuum seems at first sight to defy common sense. It asserts that every unaccelerated observer will find the speed of light in vacuum to be c, regardless of the speed of the source relative to the observer. Suppose, for example, that you are moving toward me (in vacuum) at a speed of $c/2$, and you shine a light beam at me. According to this postulate, we would both measure the speed of the light beam to be c. We know, of course, that if the light beam were replaced by a ball, we would measure different speeds for it. If you were moving toward me at 50 km/h and you threw a ball toward me at 100 km/h relative to you, I would measure its speed to be 150 km/h. Analogously, one might suppose that if you measure the speed of the light beam to be c, I should find its speed to be $3c/2$. It seems ridiculous to assert that we would both find its speed to be c. But very careful experiments verify the truth of the postulate to quite high precision; these experiments use electromagnetic radiation (visible light corresponds to a certain region of the electromagnetic spectrum) emitted by subatomic particles that have been brought, with the aid of high-energy particle accelerators, to speeds very close to c. Einstein realized that the truth of this postulate requires a careful reanalysis of our notions of space and time. When he created the special theory of relativity in 1905, he deduced what the nature of space and time would have to be in order to be consistent with the two basic postulates of the theory.

The particular part of special relativity that is relevant here has to do with the rates of clocks. Imagine yourself, equipped with an accurate clock, playing with a ball in an empty freight car on a train. The train is traveling on a straight track and is moving with constant speed v. You roll the ball across the width of the freight car, it bounces off the wall, and it rolls back to you. It is easy for you to find the average speed of the ball. The distance it traveled is twice the width of the car, and you simply divide that distance by the time it took to make the round trip across the car, as determined by your clock. Suppose now that we want to know the ball's speed relative to the ground. The distance it has traveled relative to the ground is, of course, greater than the distance relative to the train, since the train is moving. (Even if the ball were at rest on the train, it would be moving relative to the ground with just the speed of the train.) According to our everyday notions, the time it took to go across the train and back would be the same when measured on the ground as when measured on the train. We would thus find, as expected, that the ball's speed relative to the ground is greater than its speed relative to the train. Suppose now that instead of a ball you send a pulse of light across the width of the train, where it is reflected by a mirror and returned to its starting point. Just as with the ball, the distance traveled by the light pulse relative to the ground is greater than the distance it traveled relative to the train. But unlike the case of the ball, the speed of the light pulse (we suppose again that this takes place in vacuum) is the same relative to the ground as it is relative to the train. The only way this is possible, of course, is if the time it takes for the pulse to go across the train and back is greater as measured on the ground than it is as measured by your clock on the train. In other words, moving clocks run slow.

This result has nothing to do with the way clocks are built or with what happens to them as a result of being put in motion. As far as you, on the train, are concerned, your clock is keeping perfect time. But people on the ground who also have clocks that are keeping perfect time will find that your clock is running slow. Suppose that your clock shows an elapsed time T_0. Then analysis of the situation shows the corresponding elapsed time on the ground, T, to be

$$T = \frac{T_0}{\sqrt{(1 - v^2/c^2)}},$$

an expression that is known as the time-dilation formula. It follows from the principle of relativity, of course, that you would find clocks on the ground running slow, since they are moving relative to you with speed v. If a clock on the ground showed an elapsed time T_0, the corresponding elapsed time on the train, T, would be given by the same time-dilation formula.

The time-dilation formula shows that if v is close to c, a small T_0 corresponds to a large T. So if a science-fiction author wishes to get a space ship from one place to another that is, say, 10 000 light years away (i.e., it would take light, traveling with speed c, 10 000 years to get there) but wishes the occupants of the ship to age only a few years during the journey, all that is necessary is to arrange for the space ship to travel at a speed sufficiently close to the speed of light. For example, if the ship's speed, v, is close enough to c that the square root in the denominator of the time-dilation formula has the value $\frac{1}{2000}$, then during a voyage that lasts 10 000 years according to the stay-at-homes (or their descendants), the ship's passengers and crew would age only 5 years.

This is not a completely satisfactory solution to the author's problems, however. It may be desirable to have some of the voyagers return home and pick up their lives more or less as they left them. A time lapse of a decade or so may be tolerable, but if tens of thousands of years have elapsed at home the returnees may have trouble finding old friends!

It may be well to lay to rest at this point a possible worry that may have occurred to you. Let me put it in the following context: Jack and Jill, who are twins, have lived on this planet until they are both 20 years old. At that time Jill decides to go in for some high-speed travel in a space ship. Suppose that the ship's speed is so high that the time-dilation formula yields $T = 4T_0$. Jill travels on this ship for 10 years of her own time, returning when she is 30 years old. According to the time-dilation formula, Jack will have aged 40 years during her absence, and so will be 60 years old when Jill returns. You might think that, according to the principle of relativity, we could equally well consider Jack to be the traveler and Jill to be the stay-at-home, so that Jack would be younger than Jill upon her return. This would indeed be paradoxical, for when they are standing next to one another it should certainly be possible to tell simply by looking at them which one did the extra aging. (We could, of course, sharpen the paradox by letting the ship go so quickly that $T = 500T_0$. Then at the end of the journey it would indeed be possible to tell which of them had long since died of old age!)

The resolution of this apparent paradox comes from noting that the principle of relativity asserts the equivalence only of unaccelerated observers. We can suppose that it is a sufficiently good approximation to say that Jack was unaccelerated. Jill, on the other hand, must have accelerated in order to have made a round trip – at the very least, her ship would have had to change direction in order for her to return to her starting point. The theory shows that in such a situation, the time dilation formula holds when T is the time of the unaccelerated observer and T_0 that of the accelerated observer, but not vice versa. So Jill is indeed the one that is younger when she and Jack are reunited.

The science-fiction author's problem, then, is a real one. The story's characters can travel as far as they wish in as little of their own time as they wish simply by managing somehow to go at a speed sufficiently close to c, but they can't go home again – or at least they can't expect to return to a familiar world. Authors can sometimes turn this problem into an advantage, making the relativistic time-dilation effect an important part of the story line. An example of this is Joe Haldeman's novel *The Forever War*.

Still worse problems may befall the hapless travelers. Stars, galaxies, and the universe itself are all evolving. High-speed travelers may find, at the end of a long voyage, that their physical surroundings have grown much less hospitable.

Astronomical studies indicate that our sun will spend a total of about ten thousand million years stably generating energy by converting hydrogen to helium. It is believed that about half of that period has passed at the present time. At the end of that period, the sun is expected to evolve into the type of star known as a red giant, with a size that would extend well beyond the earth's orbit. Travelers returning to our solar system after the sun has evolved into a red giant would then find no earth to return to. If they wished to remain in the solar system, they would have to attempt to take up residence on some planet or other body much more distant from the sun.

The amount of time a star spends stably converting hydrogen to helium depends on the star's mass. Stars more massive than the sun evolve more rapidly, while those less massive evolve more slowly. Not just the rate, but the actual course of evolution also depends on the star's mass. Stars much more massive than the sun are expected to go through a supernova phase, in which they explode so violently that a single supernova can briefly outshine the entire galaxy in which it is imbedded. Our travelers would certainly not want to take up residence anywhere near (even on an astronomical scale) a star that might become a supernova in the foreseeable future.

As far as is known at the present, there are only three possible final ends for a star when it has exhausted all of its nuclear fuel. A star such as our sun is expected to end up as a white dwarf. In contrast to its immense size during its red-giant phase, the sun as a white dwarf will have shrunk to about the size of the earth. In this final phase of its existence, it will gradually radiate away its heat energy, ultimately becoming cold and dark. As shown by Subrahmanyan Chandrasekhar, who received the Nobel Prize for his work on these problems, no white dwarf can have a mass greater than about one-and-a-half times that of the sun. More massive stars must either shed enough mass during the course of their evolution to enable them to exist as white dwarfs or else end up in a different final state.

Some stars end up as neutron stars. Neutron stars are even more compact than white dwarfs – a neutron star with a mass equal to that of the sun would have a radius of only ten kilometers. As with white dwarfs, there is an upper limit to the mass of neutron stars. This limit is not as precisely known as is the one for white dwarfs, but it is probably in the range of a few times the mass of the sun.

The only other known final state for a star, and therefore the only known possibility for a star that ends up with more mass than is possible for a neutron star, is for it to collapse into a black hole. As a star collapses (without loss of mass) into a smaller size, its surface gravity increases. A black hole results when the collapse of a star has resulted in a surface gravity so enormous that nothing, not even light, can escape from its surface. For the sun to become a black hole, it would have to collapse into a radius of about three kilometers. The boundary of the region from which nothing can escape is known as the horizon. Anything inside the horizon is drawn ever inward by the powerful gravitational forces; for example, the star whose collapse created the black hole continues to collapse catastrophically. There is danger even outside the horizon. Although it is possible to escape from that region, a traveler venturing too close might have to accelerate away from the black hole extremely rapidly in order to avoid being drawn inside. Matter in the vicinity of a black hole can get pulled apart by the hole's strong tidal forces, and swirl around the hole as it gets drawn in. During this process, quite intense radiation can be emitted. Our space travelers certainly would not want to take up residence too close to a black hole.

Our hypothetical travelers, then, might find not only the solar system but the entire galaxy rather different from the way it was when they started on their journey. But stellar aging and death are not the only processes taking place. New stars are being formed in the galaxy, condensing out of clouds of interstellar matter. So if many regions of the galaxy have become unsuitable for the travelers, others may have come into being that would make nice places to live. Unfortunately, however, the formation of new stars cannot go on forever. Although some interstellar material is replenished through supernova explosions and other processes that take place during the course of stellar evolution, less is put back by aging stars than is needed to replace them continually with young ones. The second law of thermodynamics tells us that things will run down eventually, and the galaxy will consist only of stars in their ultimate final states, an idea that Isaac Asimov used effectively in his story *The Last Question.*

Astronomers have found that the universe as a whole is expanding. The galaxies are moving away from one another as though they were blown apart in a gigantic explosion. It is a matter of some controversy among astronomers as to whether this expansion will continue forever or whether the universe will eventually reach a maximum size and then recollapse into a fireball like the one that originally spewed it forth. In the latter case, our space travelers might face the ultimate difficulty: If they travel too long in the time frame of the universe, they may find that the whole universe is dying; there would be no hospitable places anywhere in the universe, and no place to hide from the universe's ultimate fate. This was the problem faced by the characters in Poul Anderson's novel *Tau Zero.*

Problems of this sort can arise only if the travelers' space ship reaches speeds sufficiently close to c. Even if we make the rather large assumption that travel at such speeds is technically feasible, there is still a question as to whether humans could stand the accelerations necessary to do so. An acceleration equal to the earth's gravitational acceleration, which is referred to as an acceleration of 1 g, would be quite comfortable for humans. In fact, since the acceleration would be experienced by the ship's inhabitants as though it were a gravitational field of the same magnitude but with the 'down' direction opposite to the direction of the acceleration, the ship could be constructed in such a way that the travelers would feel (gravitationally speaking, at least!) just as though they were on the earth. Accelerations of noticeably larger magnitude would probably be uncomfortable for extended periods of time, and sufficiently high accelerations would cause injury or even death. One might wonder, then, whether travel limited by such consider- ations would lead to time-dilation effects large enough to cause the worries we have considered.

Let us try to answer that question by considering a space ship traveling from the earth to some distant port of call by accelerating at 1 g (as measured in the ship's frame of reference) for half of the trip, then decelerating at 1 g for the other half. (The ship might well decelerate simply by turning around and accelerating in the opposite direction. This would not only produce the desired deceleration, but would also preserve the same sense of 'down' relative to the ship.) For the sake of simplicity, suppose also that the ship immediately returns to earth with the same mode of acceleration and de- celeration. Let us compare the time that has elapsed on earth with the amount that the travelers will have aged during this round-trip voyage.

For a voyage of only a year or two, there is not much difference between the elapsed times on earth and on the ship. In a trip of this duration, with the accelerations and decelerations limited to 1 g, not enough time is spent at speeds sufficiently close to c to make the time-dilation effect a large one. If the voyagers are gone for two years of their own time, for example, they will find upon their return that approximately twenty-five months have elapsed on earth. The effect is large enough to be clearly noticeable, but not large enough to make very much difference. If the voyage lasts for ten years of the travelers' lives, approximately twenty-five years will have elapsed on earth. This is beginning to get significant in terms of a person's life, but is still miniscule on an astronomical scale. A twenty-five year voyage for the travelers would result in a passage of more than a thousand years on earth. Although this is indeed a major discrepancy for people, it is still trivial astronomically. Even after a seventy-five year voyage for the travelers, the approximately five hundred million years that would have elapsed on earth would be only ten per cent of the expected remaining life of the sun in its present, stable phase.

Although not too many more years of the travelers' time would be required to make the elapsed time on earth greater than the sun's stable life, or even greater than the estimated present age of the universe, we are getting into times that are too long in terms of human lifetimes to be convenient for the purposes of many science-fiction stories. Furthermore, the travelers would not get very far, astronomically speaking. Only about a dozen stars lie within the distance they could reach in a ten-year (of their own time) round-trip journey of the sort we have been considering. In a forty-year round-trip journey, they could reach a distance of almost thirty thousand light years, which is a substantial fraction of the size of our own galaxy, and in a fifty-year round-trip they could reach a distance of almost four hundred thousand light years, which would enable them to visit a few of the nearby galaxies in what is known as the local group. But again, if the story line requires a lot of galaxy hopping without the characters aging considerably, this just won't do.

One common way in which authors overcome this problem is simply to postulate the existence of devices that will compensate aboard ship for the effects of acceleration. Such a device might be set to cancel out the effects of all but 1 g of acceleration, so that the occupants will feel as though they were subject to the earth's gravity even though they may be accelerating at a rate of many g's. (Depending on just what kind of device the authors are imagining into existence in their stories, the same device that compensates for high accelerations might also be used to provide 1 g of artificial gravity during periods of no acceleration.) If there are no limits to the accelerations that can be used, then it is possible to travel arbitrarily far in arbitrarily small amounts of the travelers' time, as we have seen, simply by traveling at a speed sufficiently close to c.

Although such devices free travelers from any limits on their abilities to travel, they still face the problem of coming home to vastly altered circumstances. In many science-fiction stories, it is desirable for the characters to travel to distant galaxies and find things upon their return pretty much as they left them. Ingenious authors have thus come up with other ways of having their characters move around the universe.

One such way that many authors have used is to imagine that there are certain places in the universe where a ship arriving at one of them can be immediately transported to another of them with little, if any, shipboard time spent in the passage. This is the equivalent, on a cosmic scale, of the teleportation booths that one finds in many science-fiction stories for use in virtually instantaneous transportation between two points on the earth (or some other planet). In some cases, particularly in stories of relatively recent vintage, it is supposed that black holes are the places where ships can be transported in this fashion. In order to see why black holes are favored for such travel, let us take a closer look at what theory has to say about the nature of black holes.

As we have seen, anything crossing into the interior of the horizon surrounding a black hole cannot get out again. It would seem, then, that black holes are not a good way of traveling; at least not if the travelers wish to return. But let us see what is supposed to be inside the horizon. The theory that describes such objects is Einstein's general relativity, which is the version of relativity that holds in the presence of gravitational fields. Since astronomical bodies do not seem to carry any significant amount of electric charge, let me first treat the case of an uncharged black hole. I shall then comment on how the description would be altered if the black hole carried a small charge.

The collapse of an object into a black hole may leave rapidly changing gravitational fields in its vicinity for some time. But when things finally settle down, it appears likely from the theory that there are only two possible final states for uncharged black holes; these are known as the Schwarzschild black hole, which does not rotate, and the Kerr black hole, which does. Anything crossing into the interior of the horizon of a Schwarzschild black hole must reach what is called a singularity. The exact nature of a singularity is not entirely clear – it is a place where the equations of the theory cease to hold. In the case of a Schwarzschild black hole, the singularity seems to be a place where, for example, gravitational tidal forces become infinitely strong. What it really means to talk about something becoming infinite is open to question. It may only be an indication that the gravitational field has become so intense that a better theory is needed to describe it. But at any rate, it seems clear that the neighborhood of this singularity is a decidedly uncomfortable place to be. It seems doubtful that any human observer could survive a trip close enough to the singularity to learn its true nature. Although there is also a singularity inside the horizon of a Kerr black hole, it is possible for an object to avoid running into it. By taking a suitable path, the object can emerge into a region quite like the region the object left when it originally crossed the black hole's horizon; that is, these two 'exterior' regions are described in the same way by the equations of the theory. Should the black hole carry a small electric charge, the only necessary modification of these descriptions is that in both the nonrotating and the rotating cases the object could follow a path like the one described for the Kerr black hole.

Let us, then, imagine a space ship that follows such a path between two exterior regions. Let us further imagine that the ship and its occupants somehow manage to complete the journey relatively unscathed. The theory gives no reason to think that the two exterior regions are connected to one another except through the black hole. That is, the voyagers have not taken a short cut to some place they might otherwise have reached by a much longer journey. Furthermore, they will never be able to return to the original region to tell their friends about the trip. Once they have crossed the horizon of the black hole, they cannot reemerge into the same exterior region. Authors of stories that use black holes for transportation, however, suppose that the region reached is the exterior of another black hole located somewhere in the region external to the original black hole. That is, the travelers disappear beyond the horizon of one black hole and emerge from the horizon of another one somewhere else. In doing so, they have saved themselves a much longer journey. The trouble, of course, is that things are not supposed to emerge from black holes.

A variation on this theme that might have at least a little more theoretical basis is that the ship emerges, not from another black hole, but from what is known as a white hole. A white hole is just the opposite of a black hole – things can come streaming out, but nothing can enter. Although there is no evidence that white holes might actually exist in nature, the theory allows for their existence. This, of course, is a sufficient excuse for science-fiction authors to take them seriously for the purposes of their stories.

But if we suppose that space ships can take short cuts through the universe by traveling between black holes and white holes, another problem arises. The theory then allows the possibility of the ship traveling into its own past. The sort of problem that this gives rise to is often known as the grandfather paradox: A traveler into the past meets a young man, gets into a violent argument with him, and kills him. This young man was, however, the man who would have been the traveler's grandfather had he not been thus killed. Since now the would-be grandfather never even had a chance to meet the woman who would otherwise have been the traveler's grandmother, the traveler was never born and, consequently, never traveled into the past. Therefore the grandfather, instead of being killed, did live to meet the traveler's grandmother, and so on and on around the circle. Considerations like the grandfather paradox have caused many people, including Einstein, to conclude that any solutions to the equations of general relativity that allow such travel into the past must be ruled out. Not everyone agrees with this, however, and certainly there are many science-fiction stories that explore the possibility of travel into the past.

Another favorite device of science-fiction authors for getting their characters around the universe quickly is to let them travel through 'hyperspace'. The term hyperspace is so widely used by a variety of authors that unwary readers might think there really is such a thing. If there is, it is certainly unknown to science. But science can shed light on a possible meaning of the word.

In the description of the universe provided by general relativity (as well as by most rival cosmological theories), it is possible for the three-dimensional space in which we live to be curved. In order to understand what it means to say that three-dimensional space is curved, it is helpful to consider the two-dimensional analogy. Imagine beings, somewhat like shadows, that can experience only two dimensions – that is, their universe consists of what we would call a surface, and they cannot form any intuitive notion of what it would mean to move in a direction that does not lie in the surface. (The problems of living in two dimensions have been explored in the classic science-fiction novel *Flatland*, by Edwin A. Abbott.) If their surface happened to be flat, they would find that ordinary Euclidean plane geometry held in their universe; for example, the sum of the angles in a triangle would always be 180°. But suppose that they happened to live on the surface of a large sphere. Careful measurements would show that the sum of the angles in a triangle would be greater than 180°, and – even stranger to the inhabitants of this universe – they would find that if they set out in a given direction and continued in what seemed to be a straight line, they would eventually return to their starting point.

Mathematicians among them might discover non-Euclidean geometries, and might explain to them that the geometry of their universe was that of a sphere (or, as they might call it, a hypercircle). The mathematicians might describe such a space as curved, and try to explain this notion of curvature as a higher-dimensional generalization of the concept of a curved line. The nonmathematical inhabitants of this universe might respond by saying that it makes no sense to speak of the curvature of two-dimensional space. 'Curved through what?', they might well say. 'There is no direction through which the space can curve!' The mathematicians might reply that it is not necessary to suppose the actual existence of higher dimensions; rather, it could simply be that spherical geometry was the correct geometry for their two-dimensional universe. The word 'curvature' could be thought of as simply an abstract mathematical term with no implication of physical reality for higher dimensions.

We, of course, have no trouble visualizing a sphere or any number of other curved two-dimensional surfaces. But when it comes to visualizing a curved three-dimensional space, our power to form mental images fails us (or, at least, mine does). Mathematically, it is easy to say what we mean by curvature. A flat three-dimensional space obeys three-dimensional Euclidean geometry; a curved three-dimensional space obeys some other geometry. We can give a mathematical description, for example, of a three-dimensional space that forms a hypersphere. In such a space, we would start out in a given direction and, without deviating from what seems to be a straight line, we would eventually return to our starting point. If our geometry is that of a curved three-dimensional space, this does not imply that there is any physical meaning to higher dimensions through which our space is curved. In particular, there is no implication that we could ever leave our three-dimensional space and travel through other dimensions. It would simply mean that our geometry is the appropriate non-Euclidean geometry.

When science-fiction authors write of hyperspace, it seems they are imputing physical reality to more than three dimensions. Furthermore, they seem to be supposing that our three-dimensional universe is conveniently curved through one or more higher dimensions in such a way that space travelers can save enormous amounts of time by taking a short cut through hyperspace. Returning to our two-dimensional analogy, it is as though a sheet of paper were bent into a U shape, with the distance (in three dimensions) between the two ends of the U very much less than the distance between the ends as measured along the U itself. A two-dimensional creature living on that sheet of paper could greatly shorten a voyage between the two ends by leaving the surface and taking a short cut through 'hyperspace'.

In some science-fiction stories that use hyperspace as a means of rapid transit, the authors suggest the existence of restrictions on where hyperspatial voyages can begin and end. Even in such stories, however, it is normally the case that the part of the journey that has to take place in ordinary space is not of great length. If it were, the advantages of using hyperspace in the first place would be negated. But it is hard to imagine, even mathematically, how the geometry of the universe might be arranged so that every neighborhood (astronomically speaking) is within a short hyperspatial trip of every other neighborhood. Of course, the details of the ships' hyperspatial drives are not given. Perhaps they work by, somehow, causing the three-dimensional universe to be bent in such a way that the desired places are close to one another from the point of view of hyperspace. In such a case one might worry, as science-fiction author Larry Niven suggests in his essay *Theory and Practice of Teleportation*, about what might happen if two such devices tried to operate at different places at the same time. Perhaps things would then happen in totally unexpected ways, and the poor travelers would end up thoroughly lost in a distant, unfamiliar, and hostile part of the universe. In any case, if the hyperspatial drives work by causing bending to take place, they seem to be able to cause virtually any desired bending to occur in negligible amounts of time, and to do so without other effects anywhere else in the universe; such operation is, of course, very convenient for the purposes of the stories.

Let us turn now from the question of how to travel through the universe to that of what sorts of creatures might inhabit other parts of the universe and, in particular, what we might learn from astronomy about the possibilities for life elsewhere. In some science-fiction stories, the universe seems to be populated only by our descendants; in others, there are many kinds of creatures indigenous to other locales. These creatures range from virtual copies of humans to intelligent beings existing in forms that, at first sight, we would have trouble recognizing as even being alive. One of the most interesting examples of the latter is found in the novel *The Black Cloud*, by the noted astronomer (and contributor to this volume) Fred Hoyle. The cloud of the title turns out to be a living creature of intellectual capacities far exceeding our own. In the novel, an almost plausible account is given of how such an intelligence might have come to exist in such an object.

Let us, however, not pursue the question of the existence of intelligent life forms quite different from our own; the possibilities (if we assume such life forms are indeed possible) seem to be limited only by our imaginations. If we deal instead with life not too different from our own, then we at least have some idea of the sorts of limits on physical conditions that are necessary for its existence. (For a more comprehensive treatment of the origins of life in the universe than I shall give here, see the article by C. Ponnamperuma in this volume.) There are really two distinct questions involved when we ask whether creatures similar to ourselves could exist elsewhere in the universe. The first is whether conditions elsewhere are suitable at the present time for the existence of such life. This is the sort of question that would be asked if we were concerned with populating other parts of the universe ourselves. The second is whether conditions elsewhere are, or have been, suitable for such life to have evolved indigenously.

Let us imagine ourselves planning an expedition to colonize other parts of the universe. We shall suppose that there is no difficulty in traveling wherever we want; perhaps we can reach our chosen destination by one of the means discussed in the foregoing. But certainly we shall need a planet (or at least a large asteroid – let me use 'planet' in a broad sense that would include such objects) to live on. So the first question that ought to be raised is whether planets exist around other stars.

It seems likely that the processes that led to the formation of our solar system have also operated elsewhere in the universe. Astronomers making observations in the infrared region of the spectrum have found evidence of material in orbit around other stars, although the evidence does not establish the the material is in the form of planets rather than just a diffuse dust. One way of detecting a planet would be through careful measurements of the position of a star over the course of time. If we imagine, for simplicity, a system consisting of a star and a single planet, then the two objects will orbit around their common center of mass. If the star is very much more massive than the planet, the center of mass of the system will be so close to the center of the star that the star's motion will be negligible. On the other hand, the center of mass of the system consisting of our sun and Jupiter, for example, lies just outside the sun's radius. The wobble of the star in such a system would be detectable to astronomers with sufficiently sensitive instruments in other stellar systems.

One question that has been raised about the existence of planets is whether they form in multiple-star systems. Since estimates of the fraction of apparently stellar objects that are actually multiple systems range from about one-third to practically all, it would indeed limit the possibilities of finding planets if they were likely to exist only around single stars. Certain studies have suggested that the more complicated dynamics of objects in multiple systems may lead to conditions unfavorable for the formation of planets. The situation seems far from clear, however. At least for the sake of further speculation, let us assume an abundance of planets in the universe.

In trying to find a planet to colonize, we need to be concerned about physical conditions on the planet. Without trying to be comprehensive, let me just consider some of the relevant points.

One important consideration is the planet's size. If we assume, for the sake of simplicity, that the planet is made of material with an averge density not very different from that of the earth, then the planet's surface gravity is proportional to the planet's radius. Thus if the planet is too large, the surface gravity would be uncomfortably strong. If, instead of looking for a place for us to colonize, we were looking for indigenous life, one might think that this would not be too serious a drawback; perhaps creatures could evolve under conditions of strong gravity in ways that would be adapted to those conditions. There is, however, a more subtle consideration.

It is believed that most of the material from which the solar system (and, presumably, similar systems elsewhere) formed was hydrogen. The earliest atmosphere of the earth undoubtedly had a great deal of hydrogen, although our present atmosphere contains very little free hydrogen. Some of the original hydrogen combined with other elements to form compounds; notably, of course, hydrogen and oxygen combined to form water. But most of the original hydrogen undoubtedly escaped from our atmosphere. Hydrogen is the lightest of the gases, and therefore escapes most easily. The rate of escape, however, depends on the strength of the gravitational field, and in a stronger field than the earth's, more hydrogen would be retained. Given enough free hydrogen in the atmosphere, in time it might be expected to combine with most of the free oxygen to form water, leaving an atmosphere that would be quite unsuitable for the development of life that, like our own, requires the presence of free oxygen to breathe.

On the other hand, if the planet is too small, the surface gravity would be so weak that not only hydrogen but the heavier oxygen would escape as well. (Our own moon is an example of a body too small to retain an atmosphere.) For colonization purposes, we might be able to put an artificially created atmosphere on such a planet, but we would not expect life similar to ours to evolve in an oxygen-poor atmosphere. Thus the best bet for a planet on which life similar to our own might evolve is one whose size is not too drastically different from the earth's – small enough to let hydrogen escape but large enough to retain oxygen.

Another consideration is temperature. Again, the considerations are somewhat different if we are looking for a planet to colonize than if we are asking whether life like our own could have evolved on the planet. We could certainly survive under conditions in which the external temperature is always below the freezing point of water, but if present ideas about the evolution of life on earth are correct, liquid water would seem to be a necessity for such evolution. Similarly, we wouldn't expect life like our own to evolve if the temperature is above the boiling point of water. For colonization purposes, we would find such high temperatures unhospitable, to say the least, but we might be able to find ways of sheltering ourselves sufficiently well from the environment to survive. Arthur C. Clarke, for example, has imagined humans living on the planet Mercury in his novel *Rendezvous with Rama* (in which he also imagines highly intelligent alien creatures having an interesting threefold symmetry).

The nature of the star (or stars) around which the planet orbits is one of the determinants of the temperature of the planet. Let us consider the nature of stars that are said to be on the main sequence; main-sequence stars are those in the stable 'hydrogen-burning' phases of their lives, like our own sun. (Since stars that have evolved off of the main sequence are not likely to provide a sufficiently stable environment for long-term considerations of life, I shall not include them in these considerations.) As was the case with planets, the size of the star is an important consideration. For main-sequence stars, mass, radius, and temperature are all correlated in the sense that as one increases, each of the others does also. Customarily, one talks about the mass of a main-sequence star, and its radius and temperature are then also known with reasonable accuracy.

A planet orbiting around a star significantly less massive than the sun would have to be in a smaller orbit than the earth's in order to receive enough energy from this cooler star to provide a suitable temperature range for the existence of life. But if a planet is too close to its star, tidal forces tend to cause the planet's rotation rate to become locked to its orbital period, so that the planet always has its same face toward the star. (Analogously, our moon always has its same side facing the earth as a result of tidal forces between the earth and the moon.) This would tend to make one side of the planet too hot and the other side too cold to be hospitable for life.

With stars significantly more massive than the sun, a different sort of problem arises. Because of the higher temperature of such stars, the nuclear reactions that supply their energy proceed more rapidly. Even though greater mass means there is more nuclear fuel available, the rate at which it is used increases sufficiently rapidly with mass that, as I indicated earlier, more massive stars spend less time on the main sequence than less massive stars. The most massive stars may spend less than a million years on the main sequence. Although this is very long in terms of human life times, and therefore might not be a deterrent to colonizing such a stellar system, it is a short time in terms of our evolutionary history. Thus higher forms of life might not evolve on the planets of such a star.

It seems, then, that if we are looking for places where life similar to our own might be found to have developed independently, our best bet would be to look for planetary systems around stars whose masses are not too different from the sun's mass. If we are looking for places to colonize, however, we might make do with a wider range of masses, provided that we are willing to put up with conditions somewhat different from those on earth. Of course, if we wish to worry about the really long-range future of the colonies, we may still wish to avoid the very massive stars whose stable, main-sequence lifetimes would be relatively short on the astronomical time scale.

Another temperature-related question arises when we consider the possibility of planets in multiple-star systems. A planet in such a system might sometimes be uncomfortably close to one or another of the stars, and at other times be uncomfortably far from all of them. This would make it unlikely that life would have evolved on such a planet, and it might also be an inhospitable environment for colonists. There has been a great deal of debate among theorists concerning the nature of planetary orbits in multiple-star systems, but at least some studies indicate the possibility of stable orbits that would lead to suitable temperatures, at least in systems consisting of only two stars. One case where this might happen is when the stars are well separated and the planet is essentially just in orbit around one of them, with the other causing only a small perturbation of its motion; perhaps not too different from the perturbations of the earth's motion that are caused by other planets in the solar system. The other case is when the two stars are quite close together and the planet orbits both of them at such a distance that it is almost as though there were only a single, central star.

An interesting view of life in a multiple-star system has been given by Isaac Asimov in his story *Nightfall*. The inhabitants of the planet Lagash in this story experience darkness only once every two thousand years or so, with psychological consequences that go well beyond those that some of us may have experienced in our own childhoods if we went through a period of being afraid of the dark. Asimov also makes the interesting point that in a system where orbital motions are very complicated, the familiar law of gravitation that Newton discovered might be very much harder to deduce. Perhaps, similarly, in our own world there are simple laws waiting to be discovered, whose existence is obscured by complications arising only from the way things happen to be.

A very massive star, at some point in its evolution after leaving the main sequence, undergoes a truly cataclysmic explosion in which it becomes a supernova, as I mentioned earlier. Even should one's own star be perfectly normal and safe, one would not wish to live in a part of the universe where a supernova explosion might occur relatively near by, nor is it likely that intelligent life would develop in such a region. On the other hand, supernovae have probably played an essential role in our own existence. Let me conclude this essay with a brief description of this role.

As I indicated previously, most of the universe consists of hydrogen, the lightest of the elements. It is believed that the very earliest processes in the universe, prior to the existence of the first stars, led only to the creation of hydrogen plus smaller amounts of helium, and possibly just traces of one or two other light elements. The remaining elements formed as a result of nuclear processes that take place in stars. As I have already mentioned, main-sequence stars generate energy by converting hydrogen to helium. Other stars, such as red giants, can produce energy by forming some of the heavier elements in various nuclear reactions. But this way of generating elements must stop with iron, for the production of still heavier elements absorbs energy instead of producing it. It is believed that these heavier elements are produced in supernova explosions, which provide an abundance of energy. Furthermore, these explosions spew a great deal of the star's material out into the interstellar medium. Additional material enters the interstellar medium from processes that, although less violent than super-nova explosions, nevertheless are powerful enough to eject material from the surfaces of stars.

Since stars and any associated planetary systems form by condensing out of clouds of gas and dust, the earliest stellar systems could not have contained more than trace amounts of elements heavier than helium. But as supernova explosions and other processes put heavier elements into the interstellar medium, later generations of stars and planets could contain increasingly greater fractions of heavier elements. Without elements such as carbon and oxygen, we could not exist. The idea that the material of our own bodies was born in the interior of stars distant in space and time, perhaps even in the spectacular explosions of supernovae, is one that might seem too far fetched even for science fiction, if it didn't happen to be true!